Workshop on Materials and Engineering in Aeronautics (MEA2017)

IOP Conference Series: Materials Science and Engineering
Volume 312

Moscow, Russia
15 - 16 November 2017

Editors:

Oleg N. Korsun
Evgeny V. Nikulchev

ISBN: 978-1-5108-5925-8
ISSN: 1757-8981

Printed from e-media with permission by:

Curran Associates, Inc.
57 Morehouse Lane
Red Hook, NY 12571

Some format issues inherent in the e-media version may also appear in this print version.

Printed by Curran Associates, Inc. (2018)

For permission requests, please contact the Institute of Physics
at the address below.

Institute of Physics
Dirac House, Temple Back
Bristol BS1 6BE UK

Phone: 44 1 17 929 7481
Fax: 44 1 17 920 0979

techtracking@iop.org

Additional copies of this publication are available from:

Curran Associates, Inc.
57 Morehouse Lane
Red Hook, NY 12571 USA
Phone: 845-758-0400
Fax: 845-758-2633
Email: curran@proceedings.com
Web: www.proceedings.com

Table of contents

Volume 312

2017 Workshop on Materials and Engineering in Aeronautics (MEA2017)

15–16 November 2017, Moscow, Russian Federation

Accepted papers received: 1 February 2018
Published online: 1 March 2018

IOP Conf. Series: Materials Science and Engineering **312** (2018) 011001 doi:10.1088/1757-899X/312/1/011001

PREFACE

2017 Workshop on Materials and Engineering in Aeronautics (MEA2017) was held 15-16 November 2017 in Moscow Aviation Institute (National Research University). It was patronized by Moscow Institute of Physics and Technology and Moscow Technological Institute.

This Workshop was held 13 times from 2001 to 2016. Each was attended by over 20 participants who submitted their reports for publishing. The participants represented mostly the leading technical universities of Russia. The Scientific Committee hopes that the next year the international involvement would increase.

The (MEA2017) Workshop Proceedings include 25 papers but the scope is actually huge. It covers composite materials & structure development, human-machine interface, system identification, control, robots, aircraft technical state monitoring, and that is not all. The Scientific Committee's position is not to limit the creativity of our participants. As a result all the papers are unlike one another and many of them contain original ideas and approaches. That is why our advice to a reader is not to judge the Proceedings by one or two papers but look through the most part of them – it is probable to find a touch of interesting and non-trivial.

On behalf of the Workshop Scientific Committee
the scientific editors of MEA2017 Proceedings:

Oleg N. Korsun, Dr., Prof., Head of the laboratory, State Research Institute of Aviation Systems, marmottoduo@yandex.ru

Evgeny V. Nikulchev, Dr., Prof., vice-rector Moscow Technological Institute, e_nikulchev@mti.edu.ru

IOP Conf. Series: Materials Science and Engineering **312** (2018) 011002 doi:10.1088/1757-899X/312/1/011002

Peer review statement

All papers published in this volume of *IOP Conference Series: Materials Science and Engineering* have been peer reviewed through processes administered by the proceedings Editors. Reviews were conducted by expert referees to the professional and scientific standards expected of a proceedings journal published by IOP Publishing.

IOP Conf. Series: Materials Science and Engineering **312** (2018) 012001 doi:10.1088/1757-899X/312/1/012001

Instrumental Landing Using Audio Indication

E A Burlak, A M Nabatchikov and O N Korsun*

State Research Institute of Aviation Systems, Moscow, 125319, Russia

*E-mail: marmotto@rambler.ru

Abstract. The paper proposes an audio indication method for presenting to a pilot the information regarding the relative positions of an aircraft in the tasks of precision piloting. The implementation of the method is presented, the use of such parameters of audio signal as loudness, frequency and modulation are discussed. To confirm the operability of the audio indication channel the experiments using modern aircraft simulation facility were carried out. The simulated performed the instrument landing using the proposed audio method to indicate the aircraft deviations in relation to the slide path. The results proved compatible with the simulated instrumental landings using the traditional glidescope pointers. It inspires to develop the method in order to solve other precision piloting tasks.

1. Introduction

Pilots of modern airplanes in the process of piloting have to analyze a great deal of visual information [1]. Besides, from late 70's up till nowadays, the gap between the physiological limits of a human operator and the opportunities of the technical facilities has increased considerably [1-3]. This is why the attraction of audio technologies in avionics has increased considerably.

The first experiments dealing with the audio application to the purposes of aircraft control took place in the middle of the last century [4]. From the very beginning such experiments were aimed at reducing the pilot's workload. It was assumed that a pilot is overloaded with the visual information [4], for which reason a supplementary audio channel is highly desirable.

In recent years, the growing interest to unmanned aerial vehicles (UAVs) has raised the demand for skillful operators. It is evident that the UAV operator is deprived of the sensual information, concerning accelerations, position and angular orientation of the vehicle, he cannot hear the sound, accompanying the operation of the engine, etc. Thus, the additional indication may be very useful to compensate the lack of information, associated with the remote piloting.

Within the last few years the efforts have been taken to improve the pilot cockpit's interface by means of the 3D- audio approach [5]. In this case the specially processed audio signals are fed to the pilot's left and right headphones. The signal processing is performed in such a way as to imitate the natural human perception of the direction towards the source of the sound. The experimental evaluation of this method accuracy, presented in [5], shows that it is sufficient to indicate the direction in general, but doesn't meet the requirements for precise piloting tasks. In other words, in the terms of the accuracy 3D- audio is very far from being compatible with the traditional visual indicators.

That is why in the present paper it is assumed, that in the problem of instrumental landing we should consider such distinctive and simple parameters of audio signal as loudness, frequency and modulation. The variations of these parameters are to present to the pilot the aircraft position relative to the glide path.

IOP Conf. Series: Materials Science and Engineering **312** (2018) 012001 doi:10.1088/1757-899X/312/1/012001

2. Description of the research

In order to study the audio indication channel as a source of flight information for a pilot, experiments were carried out to implement high-precision maneuvers at a flight research facility [5-8].

Experiments imitated the instrumental landing, that is the visualization of the environment outside the cockpit was turned off. Moreover the operator received information about the aircraft position relative to the glide path by audio indication channel only. To achieve this the glidescope pointers positioned on horizontal situation indicator (HSI), which normally show the aircraft deviations in relation to the glide path, were blocked and set to zero position.

Let us consider in more details the audio indication model, which we introduced specially for this research. Since the audio indication is a replacement to the HSI pointers, its inputs are connected to the unit which calculates vertical and horizontal angular deviations of the aircraft in relation to the glide path. The audio indication functions as follows. Above all, there are two "dead zones" for horizontal and vertical deviations, that is the audio channel produces no sound when both horizontal and vertical deviations are smaller than a predefined threshold. The level of this threshold was set to ±4% of the indication full scope, which in its turn reproduced the HSI glidescope pointers full range equal to ±1,6°. All this refers to both vertical and horizontal channels. The mismatches of greater level were indicated by sounds.

In this research we used three pre-prepared sound samples, which are well diverse modulated and attributed to "up", "horizon" and "bottom". The operator is to be trained a priory.

Figure 1 shows a phonogram "up" (A) "horizon" (B) and "down" (C) sounds. The figure shows that to indicate the sign of the vertical error the different sound fragment are used. To present the absolute value of the error the sound volume is used. When indicating the sign of horizontal error, sound is fed only to the left or right headphone, and the value of the horizontal error is encoded by frequency modulation of the sound. So the volume variations are reserved to the vertical channel only. In the area corresponding to the vertical "dead zone" only (when the horizon error exceeds the threshold) "horizon" fragment is used to be played at a fixed minimum volume, but the system keeps the frequency modulation for the horizon error value.

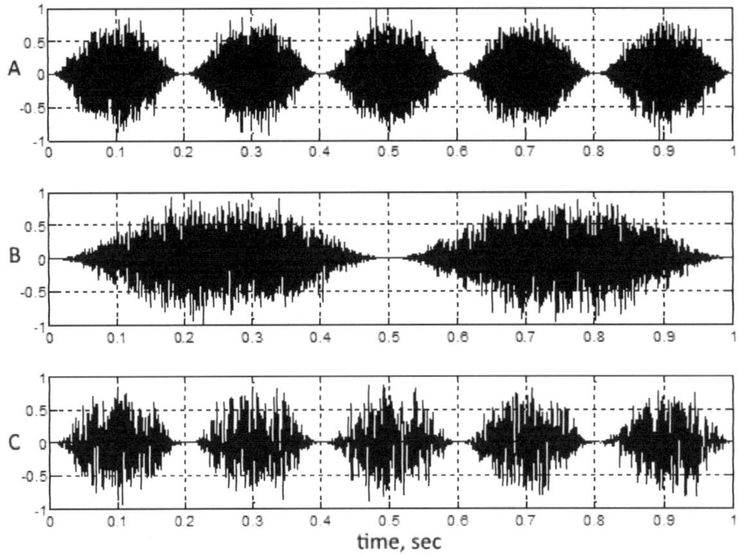

Figure 1. Phonograms of audio indication.

In the area corresponding to the horizontal "dead zone", the appropriate vertical sound is fed equally to the both headphones. So the vertical phonograms "up" (A) and "down" (C) are used in all cases except the vertical "dead zone" only, when the "horizon" (B) fragment is played.

The frequency of sound for horizontal error indication is modulated in accordance to the formula:

$$F = \frac{F_{base}}{2} + (2Amp_x - 1)P, \tag{6}$$

where F_{base} – the original sampling frequency of the audio track, Hz; F - the modulated frequency of sound, Hz; Amp_x – the normalized module of horizontal deviation; P – the scaling factor equal to 30,0 chosen according to the comments of the operator.

On figure 2 two-dimensional error domain decomposition is visualized.

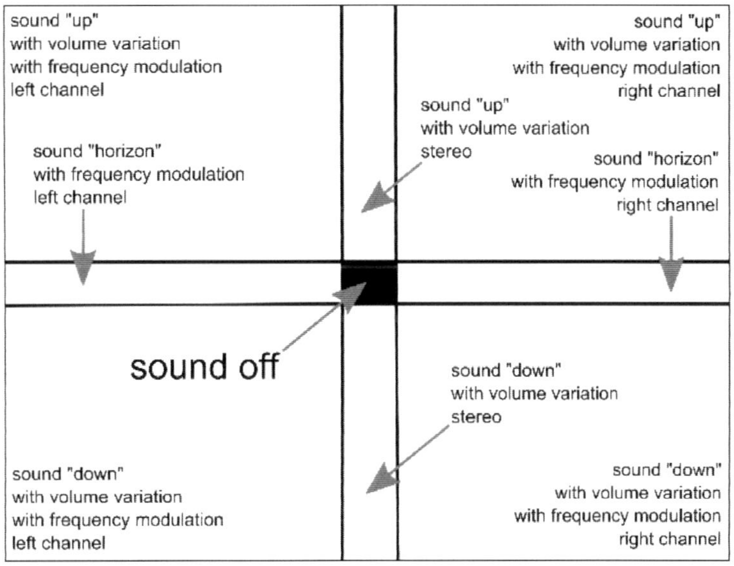

Figure 2. Two-dimensional error domain decomposition.

2. Results of the experiments

Figures 3 and 4 show the aircraft deviations in relation to the glide path obtained through the instrumental landing simulation using the glidescope pointers, while figures 5 and 6 plot the appropriate results for the landing simulation using audio indication only.

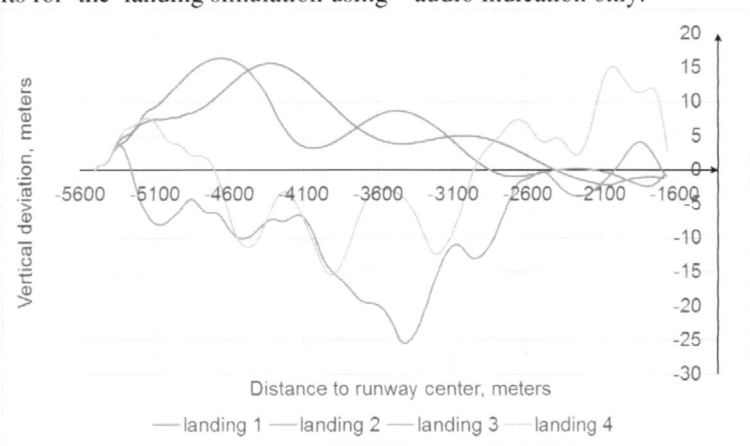

Figure 3. Vertical deviations for landing using the glidescope pointers.

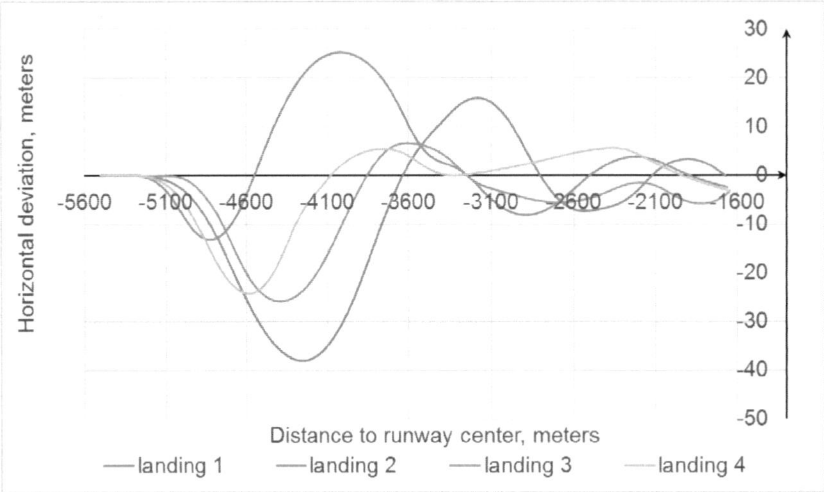

Figure 4. Horizontal deviations for landing using the glidescope pointers.

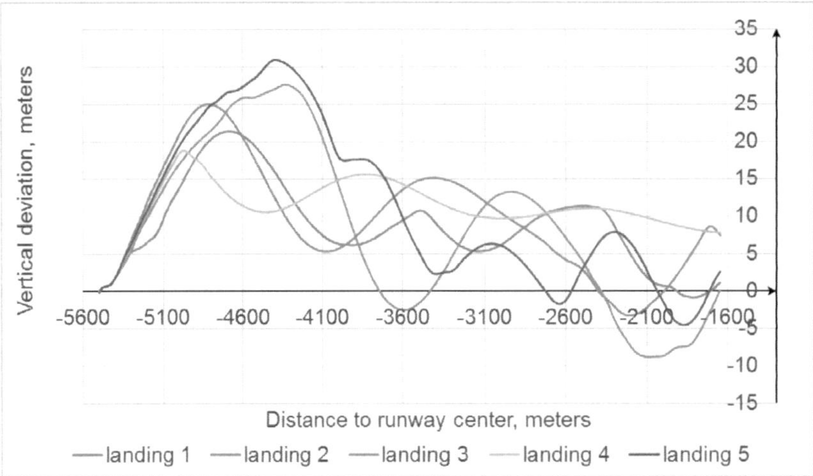

Figure 5. Vertical deviations for landing using audio indication only.

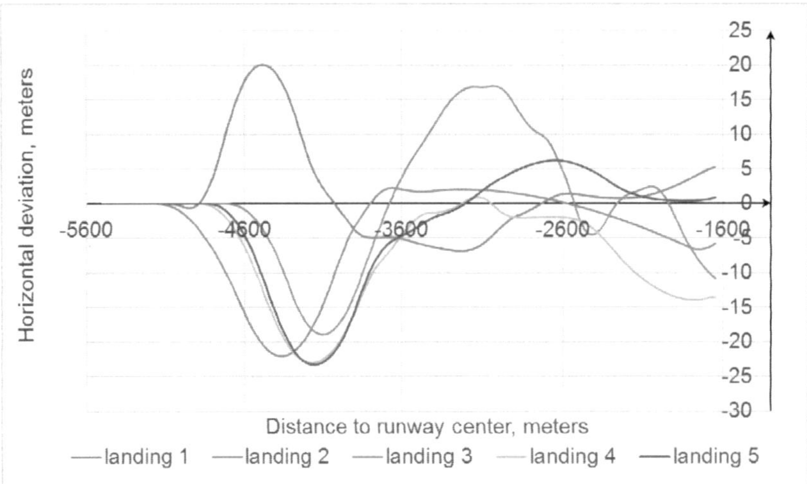

Figure 6. Horizontal deviations for landing using audio indication only.

As one can see from the figures, the results confirm the capability of the proposed audio method. Naturally the accuracy of tracking the glide path using audio indication only has deteriorated if compared to HSI glidescope pointers, but not drastically. For example, the mean misses fixed at the runway end for the landing with glidescope pointers are 1,6421 m in vertical channel and 1,0533 m in horizontal, while appropriate results for audio indication only stay within acceptable limits and make 3,18 m 7,03 m in vertical and horizontal channels respectively.

It is reasonable to investigate the audio indication method for other precision manoeuvres such as in-flight refuelling, group flight, etc.

Acknowledgments
This work was supported by Russian Foundation for Basic Research (RFBR), project 15-08-06946-a.

References

[1] Sebryakov G 2007 *Herald of Computer And Information Technologies* 10 2–7 (In Russian)
[2] Evdokimenkov V, Kim R, Krasil'shchikov M and Sebryakov G 2015 *Journal of Computer and Systems Sciences International* **54**(4) 609–620
[3] Evdokimenkov V, Kim R, Krasil'shchikov M and Sebryakov G 2016 *Lecture Notes in Computer Science, Advances in Neural Networks – ISNN 2016* **9719** (Springer-Verlag GmbH) 365–372
[4] Gerathewohl S J 1956 *Human Psychology in plane* (Moscow: Publishing House of Foreign Literature) 379 (In Russian)
[5] Lavrova G 2015 *Herald Of Computer And Information Technologies* 3 3–8 (In Russian)
[6] Burlak E and Nabatchikov A 2013 *Journal «Trudy MAI»* 68 (In Russian)
[7] Korsun O and Semenov A 2007 *Herald of Computer And Information Technologies* **7** 2–7 (In Russian)
[8] Burlak E and Nabatchikov A 2015 *Herald Of Computer And Information Technologies* **7** 47–51 (In Russian)

Adaptive fuzzy PID control for a quadrotor stabilisation

N Cherrat[1], H Boubertakh[1*] and H Arioui[2]

[1]Laboratoire d'Automatique de Jijel-LAJ, University MSBY of Jijel, Algeria
[2]IBISC-Lab, EA2456, University of Evry Val d'Essonne, France

[*]E-mail: boubert_hamid@yahoo.com

Abstract. This paper deals with the design of an adaptive fuzzy PID control law for attitude and altitude stabilization of a quadrotor despite the system model uncertainties and disturbances. To this end, a PID control with adaptive gains is used in order to approximate a virtual ideal control previously designed to achieve the predefined objective. Specifically, the control gains are estimated and adjusted by mean of a fuzzy system whose parameters are adjusted online via a gradient descent-based adaptation law. The analysis of the closed-loop system is given by the Lyapunov approach. The simulation results are presented to illustrate the efficiency of the proposed approach.

1. Introduction

Nowadays, the most famous and successful unmanned aerial vehicles (UAVs) is the quadrotor. The reasons behind this success return to its simple and useful configuration, its low cost, and its huge domain of application.

The challenge now is the development of control algorithms that ensure a reliable, safe and optimal tasks' achievement by quadrotors. Particularly, the quadrotor attitude and/or altitude stabilization is essential for many applications; it can be sufficient for some missions or a part of a control system for other ones. In fact, many classical and advanced control techniques have been applied to this end; adaptive sliding mode control [1], adaptive backstepping control [2], Lyapunov-based control [3], state feedback control [4], fuzzy PID control [5, 6] and the classical PID control[7-11].

The utilization of classical PID control is argued by its simple universal structure familiar to operators, its low cost, ease of implementation and its efficiency with a large class of systems. Unfortunately, this technique gives poor performances when the system is subject to disturbances and/or exhibits some changing characteristics. In order to deal with such circumstances, a PID control needs to be equipped with an additional mechanism for the gains online adaptation according to system and the control objective.

In this paper, one proposes an adaptive PID control scheme for the attitude and altitude stabilization of a quadrotor. Whithin this scheme, an adaptive fuzzy system for online estimation of the PID control gains such that desired performances are fulfilled.

The paper is organized as follows. In section 2 one presents the dynamical model of the quadrotor. Section 3 presents the proposed fuzzy adaptive PID control scheme. In section 4, one shows some simulation results. Section 5 concludes the paper.

2. Quadrotor model

The quadrotor is a small Unmanned Aerial Vehicle (UAV) with four propellers actuated by DC motors mounted on the end of two perpendicular arms. A basic diagram is shown in Figure 1. Each rotors pair of the same arm rotates in same direction; one pair rotates clockwise, while the other rotates counter clockwise. The quadrotor moves by adjusting the angular velocity of each rotor.

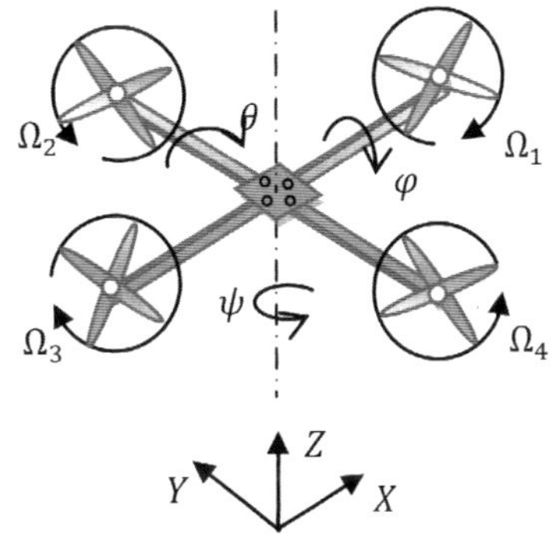

Figure 1. The quadrotor configuration [5]

Table 1. Te quadrotor model parameters

Parameter	Value
l	$0.232\ m$
d	$7.5 \times 10^{-7}\ N.m.s^2$
b	$3.13 \times 10^{-5} N.s^2$
m	$0.52\ Kg$
g	$9.806\ m/s^2$
I_x	$6.228 \times 10^{-3} Kg.m^2$
I_y	$6.228 \times 10^{-3} Kg.m^2$
I_z	$1.121 \times 10^{-2} Kg.m^2$
J_R	$6.01 \times 10^{-5} Kg.m^2$

One consider the dynamical model of quadrotor in [10] given by

$$\begin{aligned} \ddot{\phi} &= \dot{\theta}\dot{\psi}\frac{I_y - I_z}{I_x} - \dot{\theta}\frac{J_r}{Ix}\Omega_r + \frac{l}{I_x}u_1 \\ \ddot{\theta} &= \dot{\phi}\dot{\psi}\frac{I_z - I_x}{I_y} + \theta\frac{J_r}{I_y}\Omega_r + \frac{l}{I_y}u_2 \\ \ddot{\psi} &= \dot{\phi}\dot{\theta}\frac{I_x - I_y}{I_z} + \frac{1}{I_z}u_3 \\ \ddot{X} &= sin(\theta)cos(\phi)\frac{1}{m}u_4 \\ \ddot{Y} &= -sin(\phi)\frac{1}{m}u_4 \\ \ddot{Z} &= -g + cos(\theta)cos(\phi)\frac{1}{m}u_4 \end{aligned} \quad (1)$$

Where ϕ is the roll angle, θ is the pitch angle and ψ is the yaw angle. X, Y, and Z are the Cartesian position coordinates. $\Omega_r = -\Omega_1 + \Omega_2 - \Omega_3 + \Omega_4$ is the sum of rotors angular velocities. The relation between the control signal and the rotors angular velocities is given by

$$\begin{bmatrix} u_1 \\ u_2 \\ u_3 \\ u_4 \end{bmatrix} = \begin{bmatrix} 0 & -b & 0 & b \\ b & 0 & -b & 0 \\ -d & d & -d & d \\ b & b & b & b \end{bmatrix} \begin{bmatrix} \Omega_1{}^2 \\ \Omega_2{}^2 \\ \Omega_3{}^2 \\ \Omega_4{}^2 \end{bmatrix}$$

In this work we are interested by the stabilization of the attitude (angles) and the height of the quadrotor, accordingly, the quadrotor can be seen as an interconnection of four subsystems of following model form

$$\ddot{y}_i = f_i(x) + g_i(x)u_i, \; i = 1, \dots, 4 \quad (2)$$

Where $x = \left[\phi \; \dot{\phi} \; \theta \; \dot{\theta} \; \psi \; \dot{\psi} \; X \; \dot{X} \; Y \; \dot{Y} \; Z \; \dot{Z}\right]^T$ is the state vector, which is assumed available for measurement. $u = [u_1 \, u_2 \, u_3 \, u_4]^T$ is the control input vector and $y = [y_1 \, y_2 \, y_3 \, y_4]^T = [\phi \; \theta \; \psi \; Z]$ is the output vector. The nonlinear functions $f_i(x)$ and $g_i(x)$ are supposed unknown is this work.

3. Adaptive fuzzy PID control

The tracking errors are defined as

$$e_i(t) = y_{di} - y_i(t), i = 1, \dots, 4$$

and the corresponding filtered errors as

$$s_i(t) = \dot{e}_i + \lambda_i e_i, \; \lambda_i \geq 0, \; i = 1, \dots, 4 \quad (3)$$

Since the equality $s_i(t) = 0$ means that the tracking error $e_i(t)$ and all its derivatives decrease exponentially to zero [12], therefore, it is sufficient to design a control law that ensures $s_i(t) \to 0$, $i = 1, \dots, 4$.

The time derivative of the filtered tracking errors can be written as

$$\dot{s}_i = \lambda_i \dot{e}_i - f_i(x) - g_i(x)u_i \quad (4)$$

We can proove easily that the ideal control laws u_i^* given by (5) achieve the control goal

$$u_i^* = g_i(x)^{-1}\left(-f_i(x) + \lambda_i \dot{e}_i + +k_i s_i + k_{0i}\,\mathrm{sign}\left(\frac{s_i}{\varepsilon_0}\right)\right) \quad (5)$$

Where k_i and k_{0i} are positive constants and ε_0 is a small positive constant.

Since $f_i(x)$, $g_i(x)$ are unknown, the ideal controller (5) cannot be implemented. To overcome this issue and to insure the control objective, in this paper one proposes an adaptive fuzzy PID control scheme to approximate the ideal control.

In fact, each ideal control u_i^*, $i = 1,\dots,4$, is approximated by an adaptive fuzzy PID control law u_i as

$$u_i = K_{P_i} e_i(t) + K_{I_i}\int_0^t e_i(\tau)d\tau + K_{D_i}\frac{de_i(t)}{dt} = E_i^{\,T} K_{PID_i} \quad (6)$$

Where K_{P_i}, K_{I_i} and K_{D_i} are the proportional, the integral and the derivative gains respectively. $E_i = [E_{i1}, E_{i2}, E_{i3}]^T = [e_i, \int_0^t e_i(\tau)d\tau, \frac{de_i}{dt}]^T$ and $K_{PID_i} = \left[K_{P_i}, K_{I_i}, K_{D_i}\right]^T$.

Where each gains vector K_{PID_i} is estimated online using a zero order Takagi-Sugeno fuzzy system which has as input the vector E_i. For each input E_{ij} $(j = 1,2,3)$, we define m_k fuzzy sets F_{ij}^k, $k = 1,\dots,m_{ij}$, The system fuzzy rule base is the collection of $If - Then$ rules of the form

$$R_i^l: IF\ E_{i1}\ is\ G_{i1}^l\ and\ E_{i2}\ is\ G_{i2}^l\ and\ E_{i3}\ is\ G_{i3}^l\ THEN\ K_{Pi}\ is\ c_{Pi}^l\ and\ K_{Ii}\ is\ c_{Ii}^l\ and\ K_{Di}\ is\ c_{Di}^l$$

where $G_{ij}^l \in \left\{F_{ij}^1, \dots, F_{ij}^{m_{ij}}\right\}$, $l = 1,\dots,N$. c_{Pi}^l, c_{Ii}^l and c_{Di}^l are the crisp outputs of the l-th rule.

The fuzzy system output, namely the PID control gains are calculated by

$$K_{Pi} = \frac{\sum_{l=1}^N \mu_i^l c_{Pi}^l}{\sum_{l=1}^N \mu_i^l},\ K_{Ii} = \frac{\sum_{l=1}^N \mu_i^l c_{Ii}^l}{\sum_{l=1}^N \mu_i^l},\ K_{Di} = \frac{\sum_{l=1}^N \mu_i^l c_{Di}^l}{\sum_{l=1}^N \mu_i^l} \quad (7)$$

Where: $\mu_i^l = \prod_{j=1}^3 \mu_{G_{ij}^l}\left(e_{ij}\right)$, $\mu_{G_{ij}^l} \in \left\{\mu_{F_{ij}^1}, \dots, \mu_{F_{ij}^{M_{ij}}}\right\}$ and $\mu_{F_{ij}^k}(.)$ is the F_{ij}^k fuzzy membership function.

Denote $w_i^l = \frac{\mu_i^l}{\sum_{l=1}^N \mu_i^l}$, $l = 1,\dots,N$, $w_i = \left[w_i^1,\dots,w_i^N\right]^T$; $c_{Pi} = \left[c_{Pi}^1,\dots,c_{Pi}^N\right]^T$; $c_{Ii} = \left[c_{Ii}^1,\dots,c_{Ii}^N\right]^T c_{Di} = \left[c_{Di}^1,\dots,c_{Di}^N\right]^T$. From these notations (7) can be written in compact form as

$$K_{PIDi} = [K_{Pi} \quad K_{Ii} \quad K_{Di}]^T = W_i^T c_i \quad (8)$$

Where $c_i = [c_{Pi}\ c_{Ii}\ c_{Di}]^T$, and $W_i = diag[w_i, w_i, w_i]$.

Now, we replace the expression of K_{PIDi} which obtained by a fuzzy system (8) in equation (6) as

$$u_i = E_i^{\,T} W_i^T c_i \quad (9)$$

Accordingly to the universal approximation theorem [13], one assume that the fuzzy PID control laws (9) can approximate the unknown ideal controls u_i^*, which means that the unknown ideal controls u_i^* can be rewritten as

$$u_i^* = E_i^{\,T} W_i c_i^* \quad (10)$$

where $c_i^* = [c_{Pi}^{*\,T}\ c_{Ii}^{*\,T}\ c_{Di}^{*\,T}]^T$ is the unknown optimal vector value of c_i.

Now, let us find an adaptation law for the fuzzy system parameters which guarantee the error (11) between the PID control and the ideal control as small as possible.

$$e_{u_i} = u_i^* - u_i = E_i^{\,T} W_i c_i^* - E_i^{\,T} W_i c_i = E_i^{\,T} W_i \tilde{c}_i \quad (11)$$

Where $\tilde{c}_i = c_i^* - c_i$ is the vector of the parameter error.

Using (4) and (5), to define the dynamic surface as

$$\dot{s}_i = -k_i s_i - k_{0i}\,\mathrm{sign}(s_i) + g_i(x) e_{ui} \quad (12)$$

To generating the adaptation law c_i we minimize the quadratic cost function $J(c_i) = \frac{g_i(x)}{2} e_{u_i}^{\,2} = \frac{g_i(x)}{2}\left(u_i^* - E_i^{\,T} W_i c_i\right)^2$ by applying the gradient descent method on it, and introducing the so called here s-modification term $\sigma_i |s_i| c_i$ to improve the robustness performance and grantees the stability as

$$\dot{c}_i = \eta_i W_i^{\,T} E_i(\dot{s}_i + k_i s_i + k_{0i}\,\mathrm{sign}(s_i)) - \sigma_i |s_i| c_i \quad (13)$$

Unlike when we used the σ- modification or e-modification concept, the combination of the s-modification term and the $k_{0i}\, sign(s_i)$ can assure a convergence of the tracking error to zero.

Theorem 1: Consider the system (1) and assume that all the considered assumptions are verified, the control law (9) and (13) ensures that the errors $e_i(t)$ $(i = 1, \ldots, 4)$ convergence asymptotically and the boundedness of all the closed loop system signals.

Proof: To analyze the tracking errors convergence, the following Lyapunov function is consider

$$V_i = \frac{1}{2}{s_i}^2 + \frac{1}{2\eta}{\tilde{c}_i}^T\tilde{c}_i \tag{14}$$

Its time derivative is written with using (12) and (13)

$$\dot{V}_i = -k_i{s_i}^2 - k_{0i}|s_i| + s_i g_i(x)e_{ui} - {\tilde{c}_i}^T {W_i}^T E_i g_i(x)e_{ui} + \sigma_i|s_i|{\tilde{c}_i}^T c_i \tag{15}$$

By using ${\tilde{c}_i}^T c_i \leq -\frac{1}{2}\|\tilde{c}_i\|^2 + \frac{1}{2}\|{c_i}^*\|^2$, $g_i(x)s_i e_{ui} \leq \frac{g_i(x)}{4}{e_{ui}}^2 + g_i(x){s_i}^2$ and (11), (15) becomes

$$\dot{V}_i \leq -(k_i - g_i(x)){s_i}^2 - (k_{0i} - \frac{\sigma_i}{2}\|{c_i}^*\|^2)|s_i| - \frac{3g_i(x)}{4}{e_{ui}}^2 - \frac{\sigma_i}{2}|s_i|\|\tilde{c}_i\|^2 \tag{16}$$

It is clear that for $k_i \geq g_{1i}$ and $k_{0i} \geq \frac{\sigma_i}{2}\|{c_i}^*\|^2$, $\dot{V}_i$ can be bounded as follows

$$\dot{V}_i \leq -\kappa_i{s_i}^2 \tag{17}$$

Where κ_i is a positive constant

Equation (17) implies the boundedness of V_i and the parameter error vector $\tilde{c}$ (V_i, $\tilde{c}_i \in L_\infty$).

By integrating (17) we can establish that

$$\int_t^{t+T}\|s_i\|^2 \leq \frac{1}{\kappa_i}\left(V_i(t) - V_i(t+T)\right) \tag{18}$$

Since $V_i(t) \in L_\infty$, we find from (18) that $s_i \in L_2$, and we have $\dot{s}_i \in L_\infty$, consequently $s_i \in L_2 \cap L_\infty$ and $\dot{s}_i \in L_\infty$, by Barbalat's Lemma [14], we have, $lim_{t\to\infty}\|s_i(t)\| = 0$, which implies $lim_{t\to\infty} e_i = 0$.

4. Simulation results

In simulation we have take the following parameters: $\lambda_1 = 4, \lambda_2 = 3, \lambda_3 = 3,\ \lambda_4 = 2,\ k = diag[5,7,7,7]$, $k_0 = daig[20,10,11,10]$, $\sigma = 0.01, \varepsilon_0 = 0.01, \eta = 8$, the the initial states: $\varphi(0) = 0.5\,rad$, $\theta(0) = -0.5\ rad$, $\psi(0) = -0.5\,rad$, $Z(0) = 0\,m$ and the desired angles and height: $\varphi_d = 0\,rad$, $\theta_d = 0\,rad$, $\psi_d = 0\,rad$, and $Z_d = 10\,m$. For the fuzzy system, we have used the same membership functions as [15],with $\epsilon_{i1} = 1,\ \epsilon_{i2} = 0.5,\ \epsilon_{i3} = 1.5,\ i = 1, \ldots, 4$.

Figures 2-5 show the simulations results without disturbances, while figures 6-9 show the simulation results when a Gaussian noise with variance 0.004 and mean zero is added to the all the states measurements.

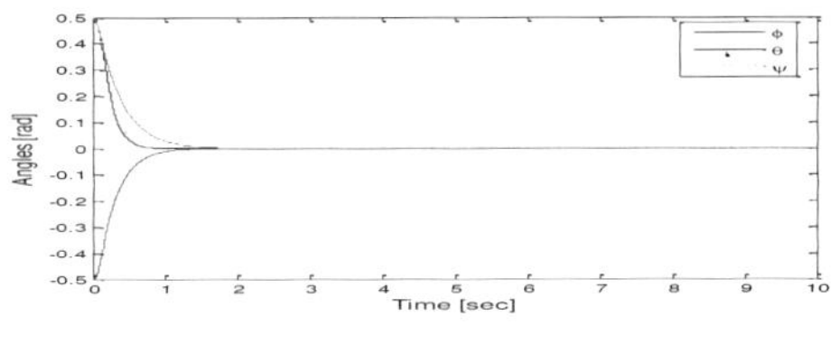

Figure 2. Angles $\boldsymbol{\theta}$, $\boldsymbol{\phi}$ and $\boldsymbol{\psi}$

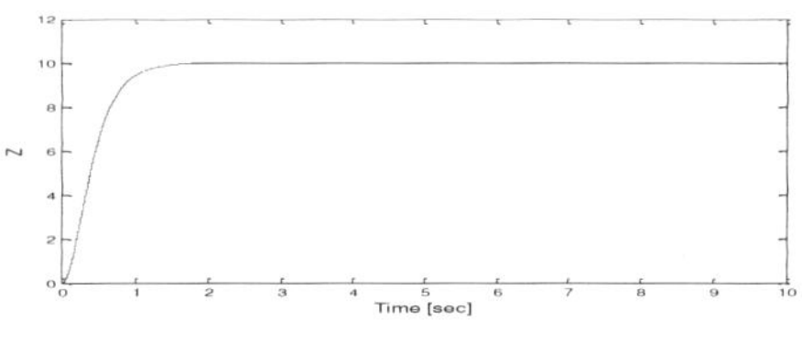

Figure 3. Height $\boldsymbol{Z}$

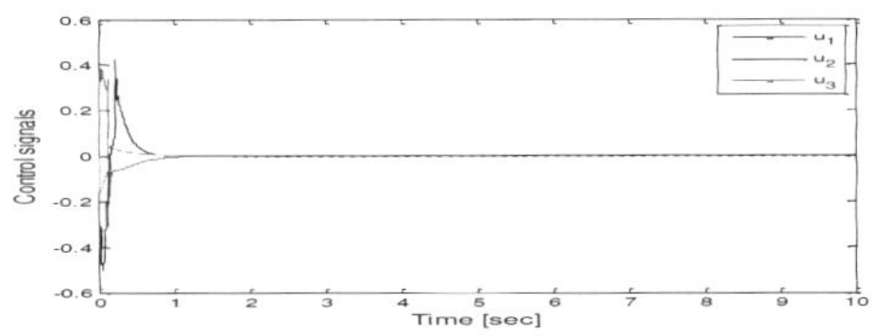

Figure 4. Control signals $\boldsymbol{u_1}$, $\boldsymbol{u_2}$and $\boldsymbol{u_3}$

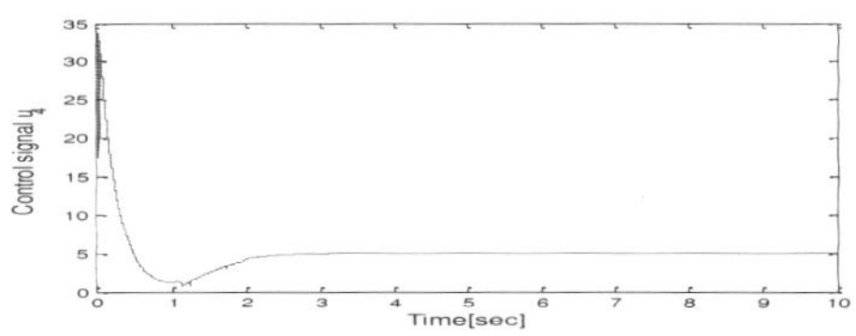

Figure 5. Control signal $\boldsymbol{u_4}$

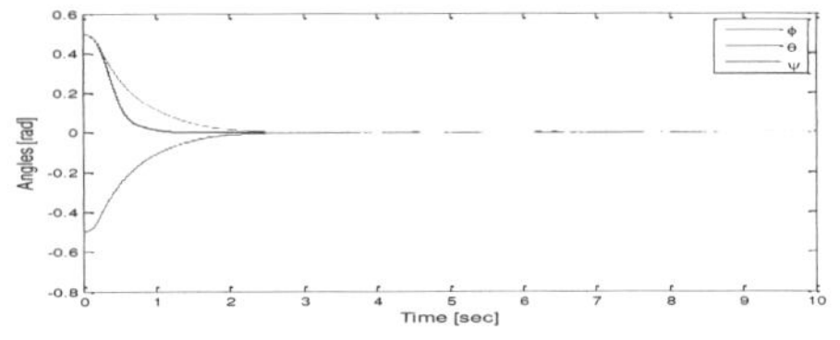

Figure 6. Angles θ, ϕ, ψ with disturbances

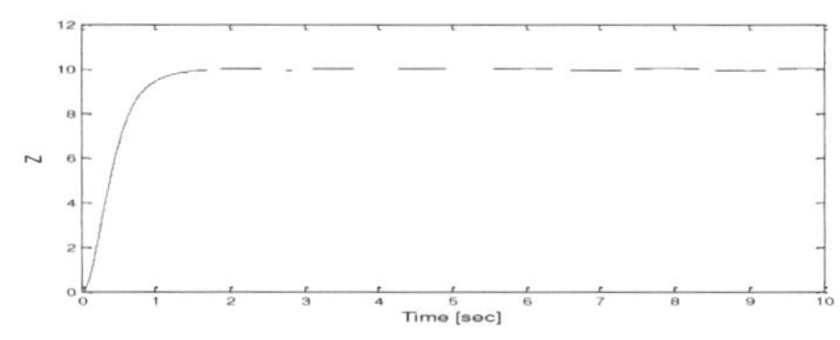

Figure 7. Height Z with disturbances

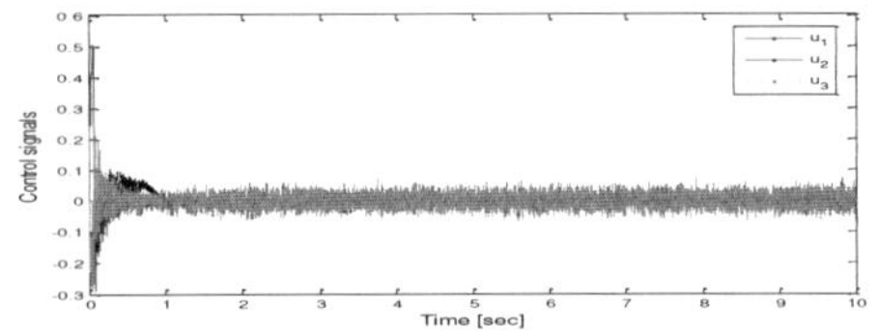

Figure 8. Control signals u_1, u_2 and u_3 with disturbances

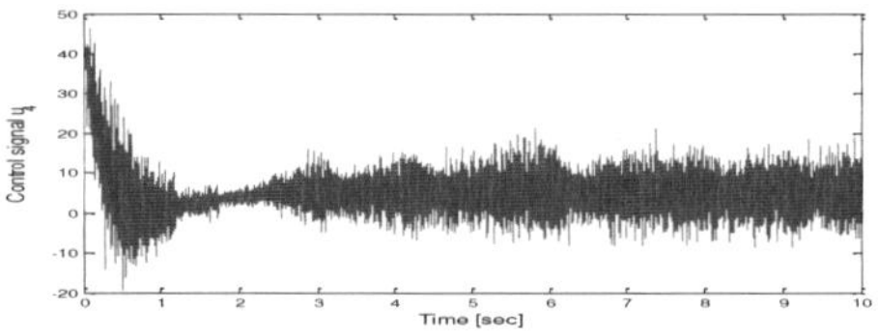

Figure 9. Control signals u_4 with disturbances

5. Conclusion

In this paper one has proposed a fuzzy PID control law for a quadrotor UAV attitude and altitude stabilization. Concretely, a PID control law to approximate an ideal control law which ensures the asymptotic stabilization of both attitude and height even with model uncertainties and disturbances. The PID control gains are estimated online by mean of an adaptive fuzzy system based on the gradient descent algorithm. The stability analysis is done using the Lyapunov approach. The simulation results show the efficiency of the proposed control scheme.

References

[1] Bouadi H, Cunha S S, Drouin A and Mora-Camino F 2011, *IEEE Inter Symposium on Computational Intelligence and Informatics (CINTI)*, 449–455.

[2] Madani T and Benallegue A 2006 Backstepping control for a quadrotor helicopter, *IEEE International Conference on Intelligent Robots & Systems*, 3255-3260.

[3] Castillo P, Dzul A, and Lozano R 2004, *IEEE Transactions on Control Systems Technology* **12** 510.

[4] Liu H, Xi J and Zhong Y 2017 *IEEE Transactions on Industrial Electronics* **64** 5585.

[5] Boubertakh H 2017 *Journal of Computer Systems Sciences International* **56** 681.

[6] Boubertakh H 2017 *MATEC Web of Conferences* **99** 03001.

[7] Bolandi H, Rezaei M and Smailzadeh M 2013, *Intelligent Control and Automation* **4** 335.

[8] Boubertakh H, Bencharef S and Labiod S 2013 PSO-based PID control design for the stabilization of a quadrotor, Algiers *Inter Conference on Systems and Control*, 514-517.

[9] Lee K U, Kim H S and Choi Y-H 2012 Hovering control for a quadrotor, *12th Conference on Control, Automation and Systems*, 162-167.

[10] Bouabdallah S 2007 *Design and Control of Quadrotors with Application to Autonomous Flying*, Ph.D. dissertation, École Polytechnique Fédérale de Lausanne.

[11] Tayebi A and McGilvray S 2006, IEEE Trans. Syst. Tech. **14** 552.

[12] Slotine J E and Li W 1991 *Applied Nonlinear Control* (New York: Wiley–Interscience)

[13] Wang L X 1994, *Adaptive fuzzy systems and control: design and stability analysis*, (Prentice-Hall, Englewood Clifs, NJ)

[14] Khalil H K 1996 *Nonlinear Systems, 2nd edn* (EnglewoodCliffs, New Jersey: Prentice-Hall)

[15] Yu W S 2006 Adaptive fuzzy PID control for nonlinear systems with Hinf tracking performance *Proc. Int. Conf. on IEEE Fuzzy Systems*, 1010-1015.

IOP Conf. Series: Materials Science and Engineering **312** (2018) 012003 doi:10.1088/1757-899X/312/1/012003

The approach to detection and application of the company's technological competences to form a business-model

A A Chursin[1], A I Kashirin[2, 3], V V Strenalyuk[2, 3], A S Semenov[2, 4 *], A A Ostrovskaya[1] and T V Kokuytseva[1]

[1] Institute of Space Technologies, People's Friendship University of Russia (RUDN University), Moscow, 17198, Russia
[2] Department of Innovative Management and External Economic Activity in Industry, the Institute of Space Technologies, People's Friendship University of Russia (RUDN University), Moscow, 17198, Russia
[3] «Rostech» State Corporation, Moscow, 119048, Russia
[4] Moscow Technological Institute, Moscow, 119334, Russia

E-mail: semenov.venture@mail.ru

Abstract The most important condition for increasing the competitiveness of business is the formation, retention, and development of key competences of the organization, which reflect the competitive advantage. This problem is especially urgent for high-tech industries, which are the most sensitive to all kinds of changes and innovations. The ways of applying the company's technological competences to form a business model, the proper form of competence description and analysis on the example of the company "Teplolux" are considered. The following from is recommended to use in IT solutions for competence databases.

1. Foreword.

The issue of applying the company's technological competences to form a business model is quite popular. On the one hand, the idea of competences is so widespread that there is no need to be an expert in any narrow field in order to become familiar with it, and many researchers offer their own approaches, classifications, methodologies, models. In the domestic practice, psychologists, teachers, sociologists, lawyers, economists and representatives of the rarest professions are engaged in the study of competences. Over the last two decades the issue of competences has been filled with a huge number of diverse studies, which, despite their academic weight and statistical justification, did not clarify the main issues from which the theory of competences began in the 1970s. Also, the way of the best competence management is still in discussions.

There is no definite answers to issue of the nature of competences, their structure and dynamics, interrelationships and interdependencies. The issue of diagnostics of competences remains open, especially in cases when it is impossible to assess, basing on the previous experience, what will be necessary in the near future. The more knowledge about competences we have, the more questions, concerning the realities, standing behind this notion arise. Therefore, the danger of any study of the problem of competence lies in the fact that it can answer some narrow, private question and leave aside the main difficulties of this theme. The authors' approach to competences is described in [1-6].

The study of the field of managerial competencies involves a special difficulty. This field excites interest for several reasons, which contributed to its popularity. Most experts believe that managerial competencies are the same in different areas of management, and differences are determined rather by position than the organization's activities. Obviously, both technological and managerial competences must be considered in business-model building tools, described in [7–9].

Managing the actions of employees is recommended for replacement with the management of characterizations of competences, and the introduction of such concepts as "strategy", "vision", "mission" gives a signal that changes in activities evaluation criteria are required. In this regard, the definition of "competences" and "competency" as special constructs which make possible to discover the content characteristics of this new quality of professional activity [10-11]. The article will deal with the application of the company's technological competencies to form a business model.

2. Introduction.

JSC "Special Design Bureau of the Cable Industry" (SDBCA) was founded in 1956. The company has key technological and marketing competences to develop and manufacture a wide range of cable products:

In 1970 - 80s the SDBCA participated in all popular aviation, space and military projects - such as 'Mars', 'Venera', 'Lunokhod', 'Salyut', 'Buran-Energia', 'Vega', ABM and SMF systems. The first optical communication line in the country, launched in Zelenograd in 1977, was made on the optical cable SDBCA. An important government order was executed for the Olympic Games 1980 in Moscow. In 1991, the socio-political and economic situation in the country has dramatically changed. The team of key employees engaged in cable electric heating starts to realize this business idea within the framework of the newly organized enterprise - LLC "Special Systems and Technologies", founded in 1991.Then the industrial production of electric underfloor heating under the brand name "Teplolyux" started. Today the group of companies "Special Systems and Technologies" (GC "SST") is the largest Russian developer and supplier of integrated solutions in the field of industrial electrical heating, including in the sphere of competence of the parent company. The total length of pipelines, which are heated by their systems, exceeds 20 thousand km. The length of super long systems of electrical heating based on the skin effect is more than 500 km. In the national rating of the Russian fast-growing technological companies (national champions) "Tekhuspekh-2016", the group of companies "Special Systems and Technologies" took the 26th place among the large companies (the revenue is over 2 billion rubles). Revenue in 2015 was 3 964 million rubles, the growth rate of the indicator for 2010-2015 is 17%, the number of employees is 1144 people.

3. Stages of formation and development of the company's business

Let's consider the stages of the company's development from the point of view of business organization in order to determine the reasons for the success of the company and evaluate the opportunity to apply positive experience to build successful businesses. By the example we show our approach to competence description in the proposed format.

3.1. Installation of an electric heating cable

At the first stage the specialists of the new company offered the installation of an electric heating cable (available in the warehouse).

This business model could be implemented by the company using the following team competences:

• general marketing competences for the organization of sales (distribution), including advertising, individual work to convince customers and enter into contracts;

• general organizational competences to ensure the work of the company, including the implementation of contracts - procurement, logistics, electrical wiring, interaction with builders, delivery of work;

• general and distinctive technological competences that ensure the core (main) activity of the company.

The general technological competence of the company is the production of electrical installation works. The team is able to perform most typical orders on electrical wiring.
The distinctive technological competence of the team in the market determines its market positioning, allocation among other electrical wiring organizations and the special sphere of activity:
- installation of an electric heating cable,
or, in a more general way:
- installation of electric "underfloor heating".
This definition of competence can be classified as a product (associated with the company's market product), and has the properties of marketing and technological competence. The properties of "marketing" show what task the customer is capable of and proposes to solve (with what tasks should be addressed - heating). The properties of "technology" is in what way the company solves the customer's task (electrical heating).
The "product" definition reflects the direction of the company's activities, but does not say anything about its competitive advantages.
Let us try to define the distinctive technological competences of the company (TC), which are components of its product competence, and provide its key activities and competitiveness. We propose the following format:

Competence No. 1:

Ability to do	Object	Object function	Supersystem	Providing the result
calculation	of cable type and size (length)	which is the resistant heating element	«underfloor heating» depending on the purpose and size of the premises	providing the necessary power of the main or additional (comfort) heating of the premises

Let us classify this technological competence.
Classification of competence No. 1:
Competence type: product, **process**
Creation stage: design, **execution or production**
Competitiveness assurance: unique, **exceptional**, standard

Competence No. 2:

Ability to do	Object	Object function	Supersystem	Providing the result
equipment configuration (selection according to characteristics and agreement)	thermostat, temperature sensor, mounting tape, coupling, mounting wire	which are the constituent elements	of the "underfloor heating"	ensuring operability

Classification of competence No. 2:
Competence type: product, **process**
Creation stage: design, **execution or production**
Competitiveness assurance: unique, **exceptional**, standard

It is possible to describe the competence for "cable muffing" and "commissioning of the system" the same way.

Possession of a complex of that data distinguishes it from a number of electrical wiring organizations and allows conducting an electrical installation business that has a greater marginality than other, simpler types of electrical installation work.

However, all the described competences are not unique (only distinctive) and therefore cannot provide the company with a stable and unconditional competitive advantage. Increased profitability of the business of installing "underfloor heating" motivates competitors to master the necessary solution. Competitive advantage of the company is determined by its preferential access to the heating cable.

The company continues to use its cable solutions for profit for some time, optimizes its expenses a little bit (for example, it starts to equip the system with its own temperature controllers and sensors, which require the corresponding solutions). Then, in an effort to expand the business in the chosen niche, goes on to the second stage of business development.

Besides "Underfloor Heating" the company offers ready-made sets for a number of heating systems:

- open areas,
- water and drain pipes,
- roof (to ensure anti-icing and snow melting).

In the market of heating systems for pipelines, the company "SST" has unique technological competences which now is assigned as core.

In the field of industrial solutions, a significant development was the development and commencement of the production of electric heating systems for super long pipes based on the SKIN effect. Serious scientific work was carried out both inside the company and in cooperation with leading research institutes. The emergence of SKIN-systems in Russia has seriously changed the vector of development of our industry, opened new prospects. "SST" is today the only Russian manufacturer of such systems and has all the necessary production and intellectual resources that allow solving the problem of electric heating with the use of SKIN-systems "on a turn-key basis". Besides "SST", the system for the electrical heating of super long pipes, based on the SKIN effect, is produced by four companies in the world, in Japan and the USA.

SKIN EFFECT is an induction-resistive system designed to maintain product temperature, protect against freezing and start-up heating of long-distance pipelines. SKIN EFFECT is the only system that allows you to heat the shoulder of the pipeline up to 30 km long with power supply from one end, without the accompanying network and the most effective and economically viable solution for heating unlimited length of main pipelines with the accompanying supply network.

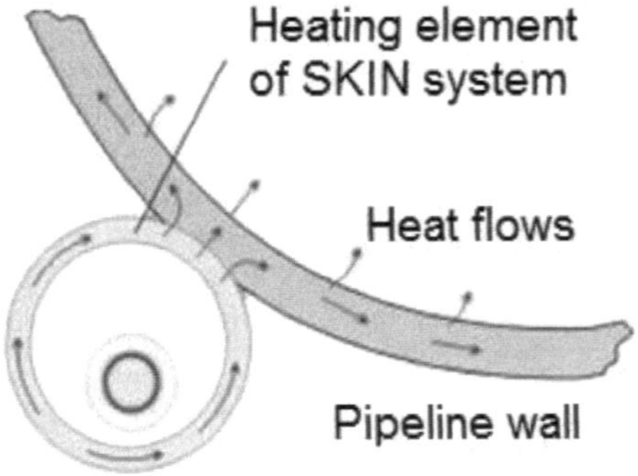

Figure 1. The principle of the SKIN-EFFECT

The SKIN SYSTEM uses special heating elements, applying the phenomenon of skin effect and the proximity effect in conductors made of ferromagnetic materials at an alternating current of the industrial frequency. The heating element is a pipe made of low-carbon steel with an external diameter of 20-60 mm and a wall thickness of at least 2 mm, inside of which there is a conductor made of a non-magnetic material (copper or aluminum) of 25-50 sq. mm section. The conductor at the end of the heating arm is securely connected to the steel pipe, and an alternating voltage is applied between the pipe and the conductor at the beginning of the arm, the value of which is calculated on the basis of the necessary heat release and the length of the heating section.

The alternating current flows through the entire section of the inner conductor, since there is no noticeable surface effect at an industrial frequency in a nonmagnetic material with good conductivity. In a ferromagnetic outer conductor (steel pipe), the skin effect is clearly pronounced and all the current flows through the inner layer of the tube with a thickness of about 1 mm, and the potential of the outer surface of the pipe remains practically zero. Due to the small thickness of the skin layer, the main heat release (up to 80%) occurs in the steel pipe.

The induction-resistive heating system (skin-system) based on the skin effect is invented in Japan (Ando Masao's patent - Ando Masao. Apparatus for maintaining liquid being transported in a pipe line at elevated temperature. Patent USA No. 3293407, published on 20.12.1966, claimed on 07.11.1963.).

This system is applied for heating long and very long pipelines (up to 6-8 km, and today - up to 15 km). By 1971, in Japan, the length of pipelines heated with the use of such devices was 120 km at the power of 15 MW. Skin-systems of electrical heating of pipelines are characterized by such positive properties as high reliability, safety, absence of the accompanying power supply network. We should note the demand for skin systems in our country, which is determined by large distances and climatic conditions. Since 2002, the company "SST" has installed and successfully used skin heating systems in dozens of facilities, both in Russia and abroad (South Korea, the United Arab Emirates). The length of super long heating systems based on the skin effect, supplied by the firm "SST" is more than 1000 kilometers.

In order to protect business, the company conducts patent work. In particular, the firm "SST" has a patent: Strupinskiy Mikhail Leonidovich. HEATED CABLE BASED ON SKIN EFFECT, HEATING DEVICE AND HEATING METHOD. Patent RU2589553C1, published on 10.07.2016, claimed on 12.03.2015.

It is interesting to note that, despite the great time spent by world manufacturers on research and development and the considerable age of the first patents, only 4 companies in the world are capable of producing such heating systems. So, we see that right and proper competence management lead to the steady development of a set of unique technological competences and keep it across time.

4. The conclusion.

The key role in building an effective innovative business is played by the distinctive "key resources" of the business model or unique competences that the "key activity" of the enterprise relies on. The competence management is the useful tool to develop the competitiveness advantages. The brief format of unique technological competence description and classification as we presented here can be used.

Acknowledgments
The article is created within the grant works financed by Russian Foundation for Humanities (order № 16-02-00711).

References

[1] Kashirin A, Semenov A, Ostrovskaya A., Kokuytseva T. 2016 *Solutions Journal of Internet Banking and Commerce* JIBC-AD 01.
[2] Kashirin A., Semenov A, Strenalyuk V, Ostrovskaya A, Kokuytseva T 2016 *International Journal of Research* **8** 50
[3] Kashirin A, Semenov A, Ostrovskaya A, Kokuytseva T, Strenaluk V 2016 *QUALITY – access to success* 17**154**

[4] Kashirin A, Semenov A, Strenalyuk V, Ostrovskaya A, Kokuytseva T 2016 *Vestnik of the Academy of Military Sciences* **4**
[5] Chesbrough H W 2003 *Open Innovation: The new imperative for creating and profiting from technology* (Boston: Harvard Business School Press)
[6] Chemezov S, Volobuev N, Koptev Yu, Kashirin A 2017 *J. Innovations* **4**222
[7] Blank S 2013 *The Four Steps to the Epiphany* (K&S Ranch Press)
[8] Chan Kim W, Mauborgne R 2015 *Blue Ocean Strategy, Expanded Edition: How to Create Uncontested Market Space and Make the Competition Irrelevant* (Harvard Business Review Press)
[9] Rodionov I, Semenov A, Gosteva E, Sokolova O 2015 *International Journal of Social, Behavioral, Educational, Economic and Management Engineering* **4** 9
[10] Nikulchev E., Ilin D and Mateshuk E 2017 *Procedia Computer Science* **103** 44
[11] Nikulchev E, Ilin D, Bubnov G and Mateshuk E 2017*International Journal of Advanced Computer Science and Applications* **8** 9

IOP Conf. Series: Materials Science and Engineering **312** (2018) 012004 doi:10.1088/1757-899X/312/1/012004

Approach to the problem of the parameters optimization of the shooting system

L A Demidova[1,2 *], **O R Gazizova**[1] **and N A Petrova**[3]

[1]Ryazan State Radio Engineering University, Ryazan, Russia
[2]Moscow Technological Institute, Moscow, Russia
[3]Branch of JSC „SRC „Progress" – Special Design Bureau „Spectr", Ryazan, Russia

*E-mail: demidova.liliya@gmail.com

Abstract. The problem of the parameters optimization of the shooting system has been considered and the possibility of application of the evolutionary algorithm for its decision has been investigated. It is offered to use the evolutionary algorithm to achieve an extremum of the fitness function under some additional restrictions. The scheme of optimization system functioning based on the evolutionary approach is resulted. The algorithm of the optimization problem decision based on the evolutionary approach is offered. Also, the idea of hybridization of the proposed evolutionary algorithm and the differential evolution algorithm is presented. This hybridization will allow reducing the time expenditures for the desired solution receiving. The example of application of the offered evolutionary approach for the problem of the parameters optimization of the shooting system for the observation object has been given.

1. Introduction

The purpose of this research is the application of evolutionary approach to the problem of parameters optimization of the shooting of the observation object. The solution of this problem is complicated by the need of the careful choice of the adequate fitness function.

This problem demands to account a large number of the restrictions for the optimization parameters values, and, also, the restriction for the fitness function value. In particular, it is necessary to account such parameters as the latitude of the point of the shooting start, the longitude of the point of the shooting start, the time of the shooting start, the shooting duration, the speed of the image motion, the azimuth and the restrictions on their values. In this case the problem of the parameters optimization belongs to the class of the multivariable one-criterion optimization problems of the parameters values, allowing reaching the greatest possible efficiency of the functioning of the shooting system.

Solving this optimization problem becomes more complicated by the fact that the initial parameters values from which it is necessary to start the search of the optimum set of the values aren't specified. A herewith, only the possible or admissible ranges of change of the parameters values are known. Therefore, we suggest to develop the software tool containing the mathematical models and the evolutionary algorithms and providing the search of the optimum parameters values of the optimization problem with the acceptable time expenditures. These found optimum parameters values must satisfy to the large number of the requirements and restrictions, specified in advance.

For the problem of the parameters optimization the use of the evolutionary approach, representing the perspective and dynamically developing direction of the intellectual data processing under the acceptable time expenditures, provides the possibility of the required decision finding with the minimum amount of the basic data.

Figure 1 shows the simulated group with 6 objects and the result of optimization of the biggest informative object. Further the aspects of the parameters' optimization of the shooting system for the chosen object will be considered.

2. The optimization system based on the evolutionary algorithm

The evolutionary approach allows applying the evolutionary algorithms (genetic algorithms, evolutionary strategies, evolutionary programming) as the optimization algorithms realizing the possibility of the simultaneous search among the several alternative solutions and choosing the best of them [1–3].

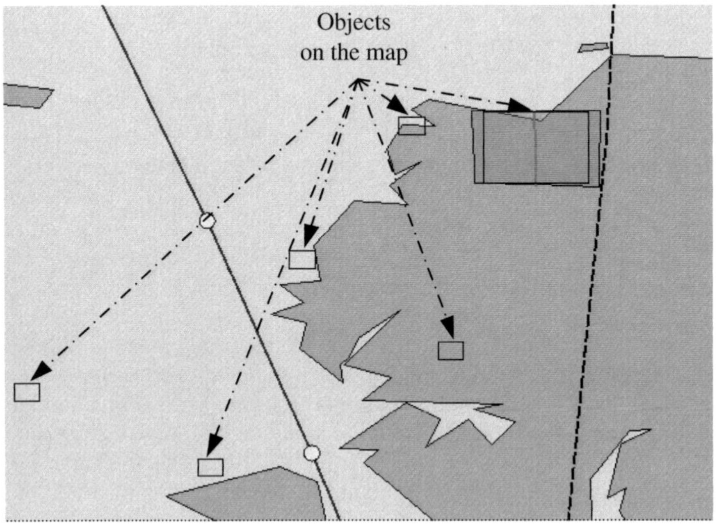

Figure 1. The group of objects and the result of optimization of the chosen object

In the case of application of the evolutionary approach it is possible to present an optimization problem in the form of the subject domain model integrated into the evolutionary algorithms. The subject domain model is understood as the representation of the set of the problem parameters, and also the criterion of its optimization in the form, acceptable for the use of the evolutionary algorithm.

The key input parameters of optimization system based on the evolutionary algorithm are the following: the number of the optimization parameters with the ranges of their possible changes, the fitness function, the population size, the number of populations, the initial values of the crossing probability and the mutation probability, the number of iterations of the evolutionary algorithm for the fitness function calculation.

The diagram of the optimization system based on the evolutionary algorithm is represented in figure 2. The connection lines between the blocks on the diagram implement the following functions:

- the connection 1 is the search of the decision on the base of the evolutionary algorithm;
- the connection 2 is the subject domain data presentation and the transfer of the received optimized parameter values by means of the optimization block;
- the connection 3 is the determination of the evolutionary algorithm search strategy;
- the connection 4 is the transfer of the optimized parameters values in the database by the evolutionary algorithm.

The software optimization system based on the evolutionary algorithm provides:

- the search of the sets of the parameters values of the optimization problem according to the subject domain model;
- the determination of the set of the "best" parameters values of the optimization problem among all decision composition received for a certain number of iterations;
- the work with the database on extraction of the initial information and the transfer of the obtained data.

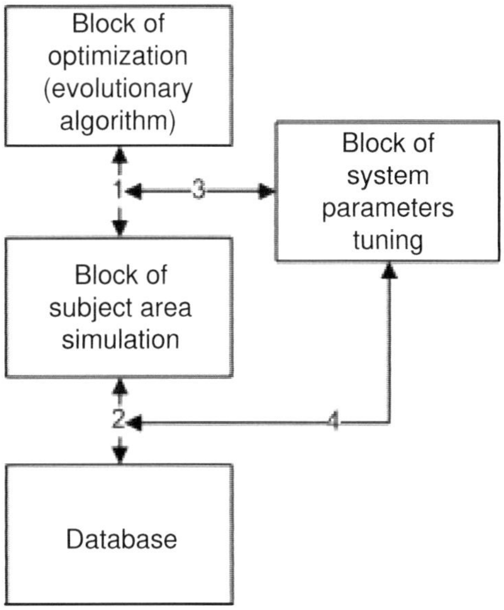

Figure 2. The diagram of the optimization system based on the evolutionary algorithm

3. The evolutionary algorithm for solving the optimization problem

The evolutionary algorithm for solving the optimization problem is represented in figure 3.

The block of the algorithm called as "The setting of the parameters and the restrictions of the optimization problem" represents a problem definition stage at which it is necessary:

- to set the variation ranges of the parameters of the optimization problem;
- to set the criterion (the fitness function) which must allow reflecting the interrelation between the problem parameters and providing the accomplishment of comparison of the received decisions versions to determine the "best" of them;
- to determine the extreme values of the parameters of the optimization problem by imposing the restrictions ("equality" or "inequality").

The block of the algorithm called as "The evolutionary algorithm implementation for the specified parameters" is responsible for implementation of the evolutionary optimization algorithm. In this block the sets of the values of the varied parameters are generated and the chromosomes after application of the evolutionary operators for the set number of iterations are determined.

The blocks of the algorithm called as "The fitness function calculation and the results ordering" and "The choice of chromosomes with the best values of the fitness function (the sets of the parameters values) for the next iteration" perform the fitness function calculation that allows comparing and choosing the "best" sets of the values of the optimized parameters with the aim to use them for the next iteration of the evolutionary algorithm. Besides, the check on the observance of all restrictions of the subject domain imposed on the values of the varied parameters and the fitness function is made here.

The block of the algorithm called as "The results display" is responsible for output of the values list of the fitness function which provide the achievement of the optimization criterion goal) and sets of the values of the optimized parameters and restrictions corresponding to this values list of the fitness function.

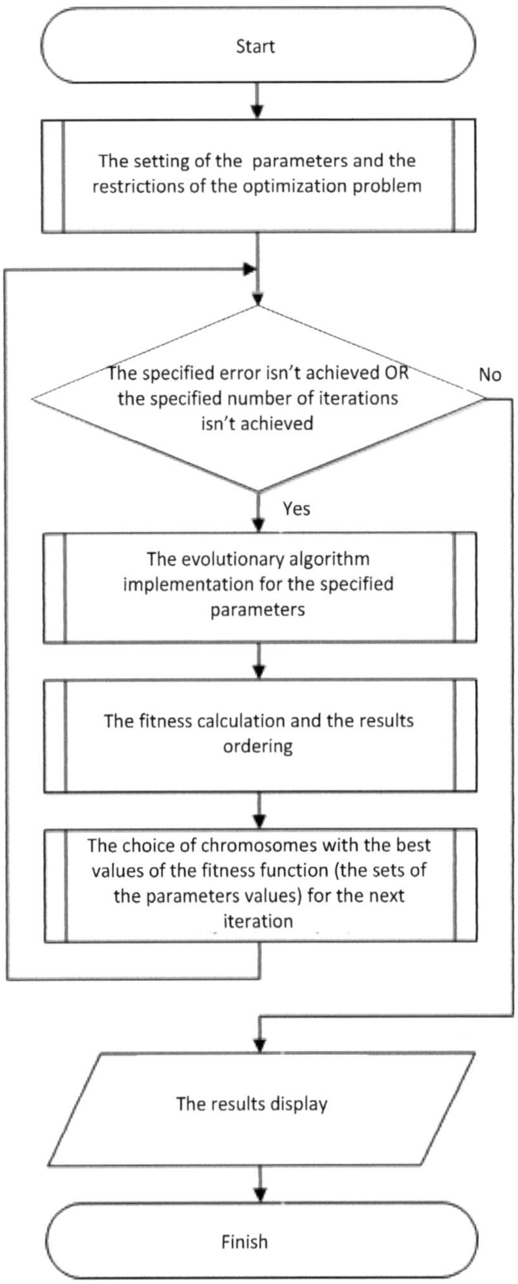

Figure 3. The evolutionary algorithm to solve the optimization problem

4. The shooting planning of the observation object

For the functional testing of the evolutionary algorithm and the software optimization system the practical optimization problem of the covering with the shooting system of the observation objects was considered.

Previously, the planning of shooting of the observation objects was made for the chosen area with the purpose of receipt of the initial parameters values of the shooting of the observation objects. The initial values of the optimized parameters of the shooting and the corresponding rangers of the permissible parameters values for one of the observation objects are given in table 1.

Table 1. The initial values of the shooting parameters and the ranges of the permissible parameters values

The shooting parameter	The initial value	The range of the permissible parameters values
The latitude of the point of the shooting start, degrees	62.5856	(61; 62.6)
The longitude of the point of the shooting start, degrees	–163.171	(–165; –162)
The time of the shooting start, second	7888174	(7888150; 7888200)
The shooting duration, second	11.8483	(5; 15)
The speed of the image motion, meters per second	60	(45; 75)
The azimuth, degrees	0	(0; 360)

The boundaries of the shooting calculated with the use of the initial values of the shooting parameters are schematically presented in figure 4 (the dark rectangle is the object; the light polygon is the shooting capture).

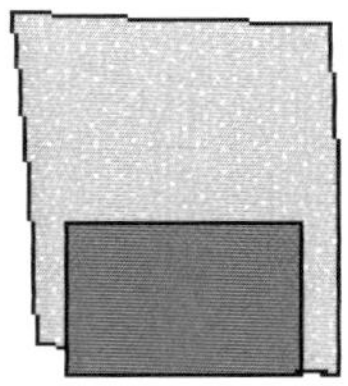

Figure 4. The boundaries of the shooting, according to the calculated initial values of the parameters

The criterion of the optimum covering of the observation object is the minimum value of a difference between the covering area and the object area. It is necessary to find the values of the shooting parameters providing the optimum covering of the observation object. To solve this optimization problem the evolutionary algorithm described above is applicable. A herewith, previously it is necessary to implement the representation of the subject domain of the optimization problem using the evolutionary algorithm terminology. The required decision of the optimization problem of the covering with the shooting of the observation object can be presented in the form of a chromosome in which the parameters, such as the latitude and the longitude of the initial shooting point, the shooting start time, the shooting duration, the image motion speed and the azimuth, which are the points in the search space, are coded.

The population with n decisions can be written down as:

$$\begin{aligned} P_1 &= (nshir_1, ndolg_1, ontel_1, dlit_1, sdi_1, azim_1), \\ &\ldots\ldots\ldots\ldots\ldots\ldots\ldots\ldots\ldots\ldots \\ P_n &= (nshir_n, ndolg_n, ontel_n, dlit_n, sdi_n, azim_n), \end{aligned} \tag{1}$$

where P_i is the i-th chromosome ($i = \overline{1, n}$); $nshir_i$ is the latitude of the point of the shooting start; $ndolg_i$ is the longitude of the point of the shooting start; $ontel_i$ is the time of the shooting start; $dlit_i$ is the shooting duration; sdi_i is the speed of the image motion; $azim_i$ is the azimuth; n is the population size.

For each gene in the chromosome P_i ($i=\overline{1,n}$) according to the restrictions imposed by the subject domain of the optimization problem the range of the permissible parameters values are defined as: $61 < nshir_i < 62{,}6$; $-165 < ndolg_i < -162$; $7888150 < ontel_i < 7888200$; $5 < dlit_i < 15$; $45 < sdi_i < 75$; $0 < azim_i < 360$.

If the latitude and the longitude of the initial shooting point, the shooting start time, the shooting duration, the image motion speed and the azimuth will be considered as the genes (the atomic elements of a genotype), then the respective set (option) of the values of the shooting parameters will represent the phenotype. Then, the aim of the evolutionary algorithm is the search of the optimum values of the shooting parameters: $nshir$, $ndolg$, $ontel$, $dlit$, sdi, $azim$.

When solving the optimization problem of the covering with the shooting system of the observation objects using the evolutionary algorithm it is expedient to apply [5]:

- the several populations of the fixed size;
- the fixed size of population corresponding to the number of the decisions of the optimization problem representing the sets of the values of the shooting parameters;
- the fixed length (digit capacity) of chromosomes, equal to six, to match the number of the optimized parameters;
- the identical combinations of strategies of selection and formation of the next generation in the each population;
- the random selection of chromosomes for the crossing;
- the single-point crossover (crossing) and the single-point mutation.

When implementing the crossing in the evolutionary algorithm, at first, the random choice of parents (two chromosomes) is carried out, then the crossing point is randomly selected and, at last, the crossing (exchange of parts) of chromosomes-parents and receiving two chromosomes-descendants is performed.

When implementing the mutation in the evolutionary algorithm, at first, the mutation point for some chromosome-parent is randomly selected, and then the mutation and receiving the chromosome-descendant is performed.

Periodically (for example, through the assigned number of iterations of the evolutionary algorithm) the accidental exchange of chromosomes between the populations is made, that allows implementing the different type of the parallel evolutionary algorithm having some properties of the island model of the genetic algorithm.

The island model assumes the existence of the several populations of the identical fixed size; the fixed digit capacity of chromosomes; the possibility to use any combinations of the selection strategies and forming of generations in populations; the absence of the crossing and the mutation restrictions; the random exchange of chromosomes between the populations.

The analysis of the subject optimization problem of the covering with the shooting system of the observation objects, shows that initially the formula for the fitness function can be written down as:

$$S_s - S_{ob} \to min, \tag{2}$$

where S_s is the shooting area, S_{ob} is the object area.

It is necessary to provide the maximum approximation of the shooting area to the area of the observation object so that vertices of the observation subject were in the borders of the optimum shooting, and the shooting area differed from the area of the observation object at the minimum value. Therefore, the specified formula for the fitness function takes the form:

$$(S_s - S_{ob}) + N \cdot N_{kr} \to min, \tag{3}$$

where S_s is the shooting area, S_{ob} is the object area; N is the number of vertices, which do not fall into the shooting boundaries; N_{kr} is the numerical value, which is much greater than the difference of areas and is required to achieve the significant deterioration in the fitness function in the case of non-compliance of condition of hit of all vertices of the object in the shooting area.

As the difference of the area of shooting and the area of the observation object will strive for the minimum value of the square kilometers, it is possible to accept the numerical value N_{kr} equal to 100000, that will allow increasing the difference of the areas of hundreds of times and will provide the rejection of decisions in which not all vertices of the object are captured.

As the shooting area must to cover the area of the observation object, it is necessary to provide the performance of the condition of positivity for the fitness function:

$$((S_s - S_{ob}) + N \cdot N_{kr}) > 0\,. \tag{4}$$

The optimization evolutionary algorithm of the covering with the shooting system of the observation objects can be described by the following sequence of steps.

1. To form M initial n-size populations with the chromosomes P_i $(i = \overline{1, n})$.

2. To make the random choice of the chromosomes parents from populations, if the current number g of iterations of the evolutionary algorithm iterations is less than the maximum number G of iterations, and then to the go to the step 3 is carried. To go to the step 6, if maximum number G of iterations is achieved.

3. To make the crossing and mutation operators in each population, if the current number g of iterations is not a multiple of some number I $(I < G)$. To choose randomly from M populations T ($T \le M/2$) pairs of populations, if the current number g of iterations is a multiple to number I, to create in the each pair of populations the pair of chromosomes with the best values of the fitness function, and to apply the crossing operation for this pair of chromosomes.

4. To form the intermediate population from the parents and the descendants. To calculate for each chromosome the value of the fitness function calculating using the formula (3) and to check the restriction (4).

5. To form the new n-size population by an exception of chromosomes with the worst values of the fitness function. To go to the step 2.

6. To choose the "best" chromosome (with the smallest value of the fitness function (3)), which defines the set of the parameters values. To output the received set of the parameters values.

When the execution of the evolutionary algorithm is over it is recommended to use the decision received by means of this algorithm (the set of the found values of the shooting parameters values) or to increase the number of iterations of the evolutionary algorithm for the further search.

The calculation with the use of the offered evolutionary algorithm of the optimum values of the shooting parameters for the observation objects for the specified set of the values of the initial parameters and specified set of the values of the restrictions was done.

The boundaries of the shooting calculated with the use of the found optimum values of the shooting parameters are schematically presented in figure 5 (the dark rectangle is the object; the light polygon is the shooting capture).

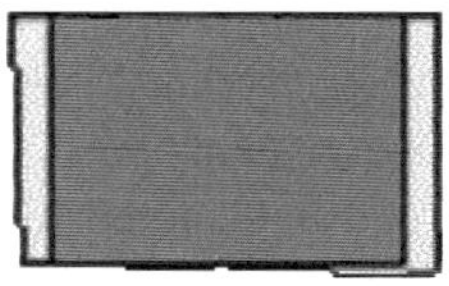

Figure 5. The boundaries of the shooting, according to the calculated optimal values of the parameters

In the figure 6 the dialog box of a program system of optimization on the basis of an evolutionary algorithm containing sets of values of the optimized parameters is shown.

The obtained results demonstrate the feasibility of application of the evolutionary algorithm for solving the optimization problem of the covering with the shooting of the observation object when the difference between the area of the object and the area of the shooting shall be minimum and the area of the shooting shall cover the area of the object with the specified number of iterations of the algorithm.

№	nshir	ndolg	ontel	dlit	sdi	azim	criteriy
1	61.816773	-163.255575	7888186.670213	9.363442	46.3462000000	148.5623520000	**418715.7100570000**
2	61.818249	-162.823117	7888157.266762	8.154576	67.5397480000	279.6447270000	**407836.2113080000**
3	62.504235	-162.480783	7888195.047214	7.304150	71.8934460000	285.4823840000	**315983.4104090000**
10	61.855858	-164.277911	7888150.000000	7.808385	75.0000000000	360.0000000000	**306721.1514520000**
25	62.514322	-163.385874	7888167.129805	7.336645	48.8520060000	129.9758580000	**304702.9382410000**
33	62.065589	-163.341501	7888158.894363	14.730092	71.7532770000	20.5389080000	**224191.2960920000**
36	62.118106	-163.041501	7888153.918413	15.000000	74.7540120000	0.0000000000	**123294.7567710000**
37	62.121650	-163.027151	7888150.000000	15.000000	75.0000000000	0.0000000000	**122673.7145140000**
64	61.738804	-163.309796	7888190.324009	14.999542	74.9999360000	0.0000000000	**40492.5442300000**
68	61.908213	-163.265190	7888184.717135	15.000000	74.2594840000	0.0000000000	**35498.0617800000**
88	62.138298	-162.898079	7888196.581075	15.000000	57.4098750000	0.0000000000	**33414.8396480000**
90	62.125523	-163.211028	7888198.104074	14.753217	53.5008270000	0.0000000000	**30958.9335550000**
95	62.103974	-163.167388	7888179.090719	15.000000	49.2356210000	7.0135780000	**20206.3263300000**
97	62.127923	-163.166698	7888179.457011	15.000000	48.2390300000	7.2525360000	**19754.8110080000**
131	61.901521	-163.195808	7888174.880735	15.000000	48.1051140000	7.0684070000	**18749.5530950000**
149	62.188046	-163.878543	7888182.335692	15.000000	45.0000000000	25.8839130000	**18657.0021780000**
152	62.306716	-163.427403	7888168.289559	15.000000	53.6290040000	15.3226360000	**17910.9289600000**
167	62.124908	-163.224806	7888181.042297	13.443936	48.6072960000	0.0000000000	**17070.0463590000**
221	61.938540	-163.100831	7888164.921802	12.721874	58.9198280000	3.4721340000	**15906.1941750000**
222	62.123653	-163.482754	7888162.606270	12.515203	57.2464630000	11.7791860000	**13923.8397460000**
224	61.982604	-163.161357	7888169.366041	14.077281	45.3399530000	0.0000000000	**13755.9515210000**
233	61.985274	-163.173268	7888169.306965	14.073842	45.4150190000	0.0065520000	**13741.5476530000**
247	62.221774	-163.129564	7888153.953973	12.913411	48.5367010000	0.0000000000	**9574.6553220000**
253	62.225932	-163.060849	7888160.699695	12.467293	45.0000000000	0.0000000000	**8815.7210630000**
258	62.304696	-163.178060	7888155.549695	12.162382	45.0000000000	0.3804950000	**7455.0224630000**
282	62.352688	-163.258505	7888150.233818	11.461740	45.0000000000	0.0000000000	**5794.5432690000**
304	62.329513	-163.246263	7888150.000000	11.247300	45.0000000000	0.0000000000	**5490.5681300000**
320	62.444091	-163.152520	7888158.627272	10.200376	45.0000000000	1.6319570000	**5191.9325600000**
360	62.429137	-163.135614	7888157.185106	9.909490	45.0000000000	1.6148180000	**4737.8583300000**
370	62.433325	-163.123473	7888156.797799	9.568921	45.0000000000	1.3670490000	**4382.0959170000**
415	62.447246	-163.134960	7888155.977683	9.218921	45.0000000000	1.2052660000	**3638.5422100000**
451	62.474317	-163.155690	7888155.001798	8.590308	45.0000000000	0.9687240000	**2883.6437040000**
485	62.491403	-163.154547	7888153.611036	8.706600	45.3608740000	1.0634660000	**2856.8779820000**
555	62.499137	-163.168288	7888154.921723	8.284917	45.0000000000	0.9485360000	**2541.2611400000**
622	62.511109	-163.182943	7888150.000000	8.366665	45.0000000000	0.6587600000	**2194.7998160000**
751	62.514427	-163.183399	7888150.000000	8.136138	45.0000000000	0.6917650000	**1891.7579870000**
875	62.514326	-163.186968	7888150.001363	8.065554	45.0000000000	0.7031500000	**1888.7377550000**
1025	62.515149	-163.187151	7888150.000000	8.075942	45.0000000000	0.7091540000	**1888.5250950000**

Figure 6. The dialog box of the program system of optimization on the basis of the evolutionary algorithm containing the sets of values of the optimized parameters

Application of the program system for optimization the values of parameters of shooting has allowed to reduce initial value of the optimization criterion (3) more, than by 221,7 times.

The execution of 1025 iterations was performed in 34 seconds using the computer on the basis of Intel Core i7 processor with a clock speed of 3.07 GHz and RAM of 1.99 GB.

Improvement of the evolutionary algorithm in the context of the shooting optimization can be executed by means of modernization of the strategy and the way of generation of the populations and descendants, and, also, by means of selection of the optimum number of populations and the optimum number of descendants. A herewith, the realization of technologies of self-adapting of the evolutionary algorithm is of considerable interest.

In addition, it is expediently to implement the hybridization of the already existing evolutionary algorithm with the differential evolution (DE) algorithm [6–8]. Nowadays, books have been published on theoretical and practical aspects of using DE in parallel computing, multiobjective optimization, constrained optimization, and the books also contain surveys of application areas. A basic variant of the DE algorithm works by having a population of candidate solutions (agents). These agents are moved around in the search-space by using simple mathematical formulae to combine the positions of existing agents from the population. If the new position of an agent is an improvement it is accepted and forms part of the population, otherwise the new position is simply discarded. This algorithm is easier to implement than the evolutionary one.

In the DE algorithm, the source of noise is not the external random number generator, but the "internal" generator, realized as the difference between randomly chosen vectors of the current population.

In case of falling into a ravine, the "cloud" of solutions takes the form of this ravine and the distribution of points becomes such that the mathematical expectation of the difference of two random vectors turns out to be directed along the long side of the ravine. This ensures the rapid movement along the narrow elongated ravines. An important feature of the DE algorithm is the ability to dynamically simulate the features of the topography of the optimized function, adjusting the distribution of the "built-in" noise source for them. This explains the ability of the DE algorithm to pass the complex ravines quickly, providing the efficiency even in the case of complex terrain. More advanced DE variants are also being developed with a popular research trend being to perturb or adapt the DE parameters during optimization. We plan to use this algorithm with the evolutionary algorithm simultaneously to find the best solution with the minimal time expenditures. In this case, one part of populations will develop on the basis of the proposed evolutionary algorithm, and the other part will develop on the basis of the DE algorithm. At the beginning, the populations will develop independently from each other for some number of generations, and then they will exchange by the best solutions. This process will be repeated until the algorithm halts. It is proposed to vary the number of populations of each type (evolutionary algorithm type and DE algorithm type) depending on their success in the current generation in the context of minimizing the criterion (3).

5. Conclusions

The developed software optimization system based on the evolutionary algorithm is the instrument of searching of the optimum parameters values of the shooting system thanks to the representation of the optimization problem in the form of model of the subject domain integrated into the evolutionary algorithm. Herewith, it is interesting to consider the possibilities of application, in particular, of the particle swarm optimization algorithm [4] in the context of solving the optimization problem as individually as in a hybrid, for example, with the genetic algorithm.

The reviewed practical example of the optimization problem of the covering with the shooting system of the observation object clearly shows the possibilities of the offered software optimization system and the prospects of its further development.

The shooting results, in particular, can be applied for the problems of the image segmentation and the objects identification [4, 9, 10].

The purpose of the further researches on application of the evolutionary approach is the development of the model of the subject domain of the problem of the optimum planning of the shooting for the objects group with the acceptable time expenditures with these or those optimization criteria of the shooting (the capture of bigger number of objects, the minimization of costs, etc.).

6. References

[1] Krus P K and Ölvander (Andersson) J 2013 *Engineering Optimization* **45**

[2] Afonin P V 2009 *Proceedings of the XV-th International Conference "Knowledge-Dialogue-Solution"* **1**.

[3] Younis A and Dong Z 2010 *Engineering Optimization* **42**.

[4] Demidova L , Nikulchev E and Sokolova Y 2016 *International Journal of Advanced Computer Science and Applications (IJACSA.)* **7** 16-24.

[5] Demidova L Petrova N and Nikulchev E 2017 *Contemporary Engineering Sciences* **10** 253-262

[6] Storn R and Price K 1997 *Journal of Global Optimization* **11**.

[7] Chakraborty U K 2008 *Advances in Differential Evolution*

[8] *Rocca P, Oliveri G and Massa A* 2011 *IEEE Antennas and Propagation Magazine* **53**
[9] Novikov A I., Yukin S A, Loginov A A, Nikiforov M B, Muratov Y R and Sablina V A 2015 *Proceedings The 5th International Workshop on Computer Science and Engineering (WCSE 2015)*
[10] Efimov A I, Novikov A I, Sablina V A 2016 *5th Mediterranean Conference on Embedded Computing (MECO) Proceedings*. Budva, Montenegro.

IOP Conf. Series: Materials Science and Engineering **312** (2018) 012005 doi:10.1088/1757-899X/312/1/012005

Approach to the problem of the parameters optimization of the shooting system

L A Demidova[1,2,*], V A Sablina[2], and Y S Sokolova[2]

[1]Moscow Technological Institute, Moscow, Russia
[2]State Radio Engineering University, Ryazan, Russia

*E-mail: demidova.liliya@gmail.com

Abstract. The problem of the objects identification on the base of their hyperspectral features has been considered. It is offered to use the SVM classifiers' ensembles, adapted to specifics of the problem of the objects identification on the base of their hyperspectral features. The results of the objects identification on the base of their hyperspectral features with using of the SVM classifiers have been presented.

1. Introduction

The problem of analysis of the hyperspectral information formed on the base of the hyperspectral images of the Earth's surface is one of the actual problems solved by the remote sensing systems [1, 2]. Also, the problem of the service development for search, indexing, cataloging and distribution of aerospace images should be solved simultaneously with the problem mentioned above [3, 4]. The russian „Resource-P" spacecrafts No. 1 – 3 with the hyperspectral equipment on the board give out the hyperspectral image (HSI) in the form of snapshots in 130 narrow adjoining ranges of visible area of spectrum. These snapshots are subjected to radiometric and atmospheric correction procedures, then the images are segmented. During the HSI processing much attention is paid to the questions of the objects identification on the base of their hyperspectral features. The datasets of the hyperspectral features (HSF) is formed on the basis of the segmented images.

Other paragraphs are indented (BodytextIndented style).

Every dataset is the text file containing information on the wavelength of the recorded radiation and the spectral reflection coefficient (SRC) (or the spectral brightness coefficient (SBC)) of the object. However, at the solution of the problem of the objects identification on the base of their HSF the use of the SRC dependence on wavelength is more preferable because the brightness feature doesn't depend on the shooting conditions in such degree as the SBC. Besides, unlike the SBC, for obtaining the SRC values the standard reflecting surface in sight of the analyzed object isn't necessary.

The HSF of the earth's surface object can be represented graphically in the form of a spectral curve reflecting the relationship between the wavelength and the SRC values of the analyzed object [1].

Figure 1 shows the main stages of the HSF obtaining of the analyzed object based on the HSI. Also, figure 1 shows the fragment of the text file with the object's HSF.

The Support Vector Machine (SVM) algorithm is the supervised machine learning algorithm. This algorithm is one of the boundary classification algorithms [5, 6]. Nowadays, it is used to solve different classification problems in various applications with great success.

The SVM classifier is used for training, testing, and classification. Satisfactory quality of training and testing allows using the SVM classifier to classify new objects. Choosing optimal parameters' values for the SMV classifier is a very significant problem. It is necessary to find the kernel function type, values of the kernel function parameters and value of the regularization parameter [5, 6]. The found optimal parameters' values participate in the formation of the classifying function $F(x)$, which compares the object to the class with the label from the set $\{-1;\ +1\}$. It is difficult to provide implementing of high-accuracy data classification with the use of the SVM classifier without adequate solution to this problem.

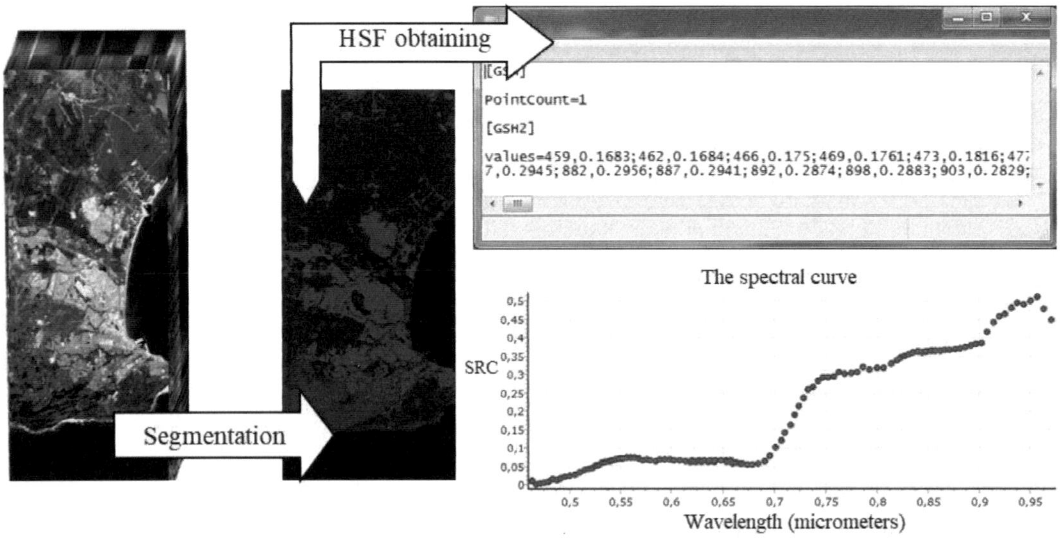

Figure 1. The main stages of the HSF obtaining for the analyzed object

Nowadays, the interest in the issue of increasing the accuracy of classifiers based on the machine learning algorithms by combining the capabilities of several classifiers and creating the classifiers' ensembles has increased significantly [7 – 9]. The training of the classifiers' ensemble is understood as the procedure of training of the final set of the private classifiers, whose private solutions are then combined to form the resulting classification decision on the basis of the aggregated classifier.

We propose to use the SVM classifiers' ensembles, formed on the basis of various strategies of the private classifiers' integration, to solve the identification problems of the objects on the basis of their HSF according to the HSI in order to improve the classification accuracy.

2. SVM algorithm

Let the experimental dataset be a set in the form of $\{(z_1, y_1),\ ...,\ (z_s, y_s)\}$, in which each object $z_i \in Z$ ($i = \overline{1, s}$; s is the number of objects) is assigned to a number $y_i \in Y = \{+1;\ -1\}$ having a value of +1 or −1 depending on the class of the object z_i. It is assumed that every object z_i is mapped to q-dimensional vector of numerical values of features i (typically normalized by values from the interval [0, 1]) where z_i^l is the numeric value of the l-th feature for the i-th object ($i = \overline{1, s}$; $l = \overline{1, q}$) [5, 7]. It is necessary to use the kernel function $\kappa(z_i, z_\tau)$ to build the classifier $F : Z \to Y$, which compares the class to the number from the set $Y = \{+1;\ -1\}$ or some object from the set Z. In training of the SVM classifier it is necessary to determine the kernel function type $\kappa(z_i, z_\tau)$, values of the kernel parameters and value of the regularization parameter C, which allows finding a compromise between maximizing of the gap separating the classes and minimizing of the total error [5 – 9]. A herewith typically one of the following functions is used as the kernel function $\kappa(z_i, z_\tau)$: linear function; polynomial function; radial basis function; sigmoid function [5, 6].

To build "the best" SVM classifier it is necessary to implement the numerous repeated training (for the training data set with S elements) and testing (for the test data set $s-S$ elements, $S<s$) on the different randomly generated training and test sets with following determination of the best SVM classifier in terms of the highest possible classification quality provision. The SVM classifier with satisfactory training and testing results can be used to classify new objects [5 – 9].

As a result of the training, the classification function is determined in the following form [5 – 7]:

$$f(z) = \sum_{i=1}^{S} \lambda_i \cdot y_i \cdot \kappa(z_i, z) + b\,. \tag{1}$$

The classification decision, associating the object z to the class -1 or $+1$, is adopted in accordance with the rule [5 – 7]:

$$F(z) = sign((z)) = sign\left(\sum_{i=1}^{S} \lambda_i \cdot y_i \cdot \kappa(z_i, z) + b \right). \tag{2}$$

3. The SVM ensembles

In most cases SVM classifier provides high quality of data classification. In exceptional cases the SVM ensembles can be used to increase the classification accuracy. The using of the SVM ensemble allows fulfilling the high-precision data classification, especially Big Data classification, with the acceptable time expenditures.

After training, each classifier generates its own (private) classification decisions, same or different from the actual results of classification. Accordingly, the different private SVM classifiers correspond to the different classification accuracy. The quality of the received classification decisions can be improved on the base of ensembles of the SVM classifiers [7 – 9]. In this case, the finite set of privately trained classifiers must be learned. Then the classification decisions of these classifiers are combined. The resulting solution is based on the aggregated classifier. The majority vote method and the vote method based on the degree of reliability can be used as the rules (strategies) of the definition of the aggregated solutions.

The majority vote method is one of the most common and frequently used methods for combining decisions in the ensemble of classifiers. But this method does not fully use the information about the reliability of each private SVM classifier. For example, suppose that the SVM classifier ensemble aggregates the results of five private SVM classifiers, where values of the function $f(z)$ of the object z (1) obtained from the three private SVM classifiers, are negative (class –1), but very close to the neutral position, and values of the function $f(z)$ of the other two SVM classifiers are strongly positive (class +1), i.e. very far away from the neutral position. Then the result of the aggregated decision of the ensemble on the basis of "one classifier – one vote" is following: the object z belongs to the negative class (majority vote), although it is obvious, that the best and more appropriate choice for the object z is a positive class. Despite the good potential of the majority vote method for combining of the group of decisions, it is recommended to use other methods to increase the accuracy of classification.

Vote method based on the degree of reliability uses value of the function $f(z)$ for the object z obtained by each private SVM classifier. The greater the positive value of $f(z)$ in (1) returned by the SVM classifier, the more precisely the object z is determined in class +1, and the less negative value $f(z)$, the more precisely the object z is defined in class –1. Values "–1" and "+1" for $f(z)$ indicate that the object z is situated on the boundary of the negative and positive classes, respectively.

When using an ensemble of classifiers for solving classification problems special attention should be paid to the methods of forming a set of private classifiers, which can later be used in the development of the final SVM classifier. It is experimentally confirmed [7 – 9], that the ensemble of classifiers shows better accuracy than any of its private members, if private classifiers are accurate and varied. Therefore, the formation of the set of the private SVM classifiers is required: 1) to use the various kernel functions; 2) to build classifiers in the different ranges of change of the kernel parameters and regularization parameter; 3) to use various sets of training and test data. To select the appropriate members of the ensemble in the set of the trained SVM classifiers it is recommended to use the principle of maximum decorrelation. In this case the correlation between the selected classifications should be as small as possible. After training, each private j-th classifier from the k trained classifier will correspond to a certain array of errors: $e_{ij} = | y_{ij} - \tilde{y}_{ij} |$, where e_{ij} is the error of j-th classifier at i-th row of the experimental data set ($i = \overline{1,s}$; $j = \overline{1,k}$); y_{ij} is the classification decision (−1 or +1) of j-th classifier at i-th row of the experimental data set; y_{ij} is the real meaning of a class (−1 or +1), for which the i-th object is belong to.

The SVM classifiers not permitting an error on the experimental data set should be excluded from further consideration and from the remaining quantity of the SVM classifiers. It is necessary to select an appropriate number of private SVM classifiers with maximal variety. To solve this problem decorrelation maximization algorithm can be used. This algorithm provides a variety of private SVM classifiers, being used in the construction of the ensemble [7]. If the correlation between the selected classifiers is small, then the decorrelation is maximal.

Let there be an error matrix E of set of private SVM classifiers with size $s \times k$:

$$E = \begin{bmatrix} e_{11} & e_{12} & \dots & e_{1k} \\ e_{21} & e_{22} & \dots & e_{2k} \\ \dots & \dots & \dots & \dots \\ e_{s1} & e_{s2} & \dots & e_{sk} \end{bmatrix}, \quad (3)$$

where e_{ij} is the error of the j-th classifier at the i-th row of the experimental data set ($i = \overline{1,s}$; $j = \overline{1,k}$).

On the basis of the error matrix E (3) the following assessments can be calculated [7]:

- mean:

$$\bar{e}_j = \frac{1}{S}\sum_{i=1}^{s} e_{ij} \quad (j = \overline{1,k}); \quad (4)$$

- variance:

$$V_{jj} = \frac{1}{S}\sum_{i=1}^{s} (e_{ij} - \bar{e}_j)^2 \quad (j = \overline{1,k}); \quad (5)$$

- covariance:

$$V_{tj} = \frac{1}{S}\sum_{i=1}^{s} (e_{ij} - \bar{e}_j)\cdot(e_{it} - \bar{e}_t) \quad (j = \overline{1,k};\ t = \overline{1,k}); \quad (6)$$

Then the elements r_{tj} of the correlation matrix with size $k \times k$ are calculated as:

$$r_{tj} = V_{tj} \Big/ \sqrt{V_{tt} \cdot V_{jj}}; \quad (7)$$

where r_{tj} is the correlation coefficient, representing the degree of correlation of t-th and j-th classifiers ($j = \overline{1,k}$; $t = \overline{1,k}$); $r_{jj} = 1$ ($j = \overline{1,k}$).

Using the correlation matrix R it is possible for each private j-th classifier to calculate the plural-correlation coefficient ρ_j, which characterizes the degree of correlation of the j-th and all other $(k-1)$ classifiers with numbers t ($t = \overline{1,k}$; $t \neq j$) [7]:

$$\rho_j = \sqrt{1 - |R| / R_{jj}} \quad (j = \overline{1,k}), \quad (8)$$

where $|R|$ is the determinant of the correlation matrix R; R_{jj} is the cofactor of the element r_{jj} of the correlation matrix R.

A quantity ρ_j^2 called the coefficient of determination. It shows the proportion of the variation of the analyzed variable, which is explained by variation of the other variables. The coefficient of determination ρ_j^2 can take values from 0 to 1. The closer the coefficient to 1, the stronger the relationship between the analyzed variables (in this case, between private classifiers) [7]. It is believed that there is a dependency, if the coefficient of determination is not less than 0.5. If the coefficient of determination greater than 0.8, it is assumed that high dependence exists.

For selection of private SVM classifiers for integration into the ensemble it is necessary to determine the threshold θ. Thus, the j-th private classifier must be removed from the list of classifiers if the coefficient of determination ρ_j^2 satisfies to condition $\rho_j^2 > \theta$ ($j = \overline{1,k}$). If it is necessary to identify the most various classifiers, generating decisions with the most different arrays of errors on the experimental data set, thresholds θ, satisfying to condition $\theta < 0.7$ should be selected. The additional considerations can be also taken into account to avoid the exclusion of insufficient or excessive number of private SVM classifiers.

The decorrelation maximization algorithm can be summarized into the following steps [7].

Step 1. To calculate the matrix V and the correlation matrix R with formulas (5), (6) and (7) respectively.

Step 2. To calculate the multiple correlation coefficients ρ_j ($j = \overline{1,k}$) with (18) for all classifiers.

Step 3. To remove classifiers, for which $\rho_j^2 > \theta$ ($j = \overline{1,k}$), from the list of classifiers.

Step 4. To repeat iteratively steps 1 – 3 for the remaining classifiers in the list until for all classifiers the condition $\rho_j^2 \le \theta$ ($j = \overline{1,k}$) will not satisfied.

As a result, the list of classifiers used to form the ensemble will consist of m ($m \le k$) private classifiers.

For classifiers selected in the ensemble, it is necessary to carry out:

- the rationing of degrees of the reliability;
- the strategy search for the integration of members of the ensemble;
- the calculation of the aggregated decision of the ensemble.

Value of the reliability $f_j(z)$, which is defined for the object z by the j-th classifier, falls into the interval (–∞, +∞). The main drawback of such values is that in the ensemble the private classifiers with large absolute value are often dominated in the final decision of the ensemble. To overcome this drawback, the rationing is carried out: the transformation of values of degrees of reliability in the interval [0; 1] is fulfilled. In the case of binary classification in the rationalization for the object z the values of the reliability of its membership to positive class (labeled +1) $g_j^+(z)$ and to negative class $g_j^-(z)$ are determined. These values can be determined by the formulas [7]:

$$g_j^+(z) = \frac{1}{1+e^{-f_j(z)}}, \tag{9}$$

$$g_j^-(z) = 1 - g_j^+(z). \tag{10}$$

The selected private classifiers are combined into the ensemble using $g_j^+(z)$ and $g_j^-(z)$ ($j = \overline{1,m}$) in accordance with one of the following five strategies [7].

1. Maximum strategy:

$$A(z)=\begin{cases}1, & \text{if } \max_{j=1,m} g_j^+(z)\geq \max_{j=1,m} g_j^-(z),\\ -1, & \text{otherwise.}\end{cases} \quad (11)$$

2. Minimum strategy:

$$A(z)=\begin{cases}1, & \text{if } \min_{j=1,m} g_j^+(z)\geq \min_{j=1,m} g_j^-(z),\\ -1, & \text{otherwise.}\end{cases} \quad (12)$$

3. Median strategy:

$$A(z)=\begin{cases}1, & \text{if } \frac{1}{m}\sum_{j=1}^{m} g_j^+(z)\geq \frac{1}{m}\sum_{j=1}^{m} g_j^-(z),\\ -1, & \text{otherwise.}\end{cases} \quad (13)$$

4. Mean strategy:

$$A(z)=\begin{cases}1, & \textit{if } \sum_{j=1}^{m} g_j^+(z)\geq \sum_{j=1}^{m} g_j^-(z),\\ -1, & \text{otherwise.}\end{cases} \quad (14)$$

5. Product strategy:

$$A(z)=\begin{cases}1, & \text{if } \prod_{j=1}^{m} g_j^+(z)\geq \prod_{j=1}^{m} g_j^-(z),\\ -1, & \text{otherwise.}\end{cases} \quad (15)$$

The value $A(z)$ is an aggregated measure of the reliability's value of the SVM classifier ensemble. It can be used to integrate the members of the ensemble [7].

The learning algorithm of the ensemble of the SVM classifiers can be summarized into the following steps.

Step 1. To divide the experimental data set into k training data sets: TR_1, ..., TR_k.

Step 2. To learn k private SVM classifiers with the different training data sets TR_1, ..., TR_k and to obtain k private SVM classifiers (ensemble members).

Step 3. To select m ($m \leq k$) SVM classifiers from k classifiers using the decorrelation maximization algorithm.

Step 4. To determine values of m classification functions for each selected private SVM classifier: $f_1(z)$,..., $f_m(z)$.

Step 5. To transform values of degrees of reliability, using (19) and (20), for the positive class $g_1^+(z)$,..., $g_m^+(z)$ and for the negative class $g_1^-(z)$, ... , $g_m^-(z)$.

Step 6. To determine the aggregated value $A(z)$ of the reliability of the SVM classifier ensemble using (11) – (15).

This algorithm, used for the weak SVM classifiers, will provide a better quality of the classification accuracy than accuracy of any single private classifier used for aggregation.

The problem of choosing of the threshold θ is very important. Value θ for which all five rules of classification (11) – (15) show stable improvement of the classification quality must be chosen as the threshold value θ^* ($\theta^* < 0.7$). Thus the use of each of the five rules leads to improvement of the classification quality resulting in the reduction of the number of erroneous decisions, when the smaller number of private classifiers, corresponding to the threshold value θ^*, is applied. Such stable $\theta' > \theta^*$ improvement of the classification quality isn't observed for all examined values θ' (for which).

It should be noted, that the majority vote rule may be used for decisions, obtained using the classification rules (11) – (15), to determine the required threshold value θ^*.

4. Experimental studies

The proposed algorithms were combined into the united software named as "Intellectual Classification", which allows fulfilling the intellectual classification of objects using the SVM algorithm and providing the ability of combining of the several private classifiers into an ensemble to improve the classification quality.

In the course of research, the identification of objects according to the hyperspectral shooting data was made. In particular, the detection problem of water objects and anthropogenic ones was solved.

The base of spectral standards containing 220 HSFs of natural and artificial objects was formed on the basis of the real hyperspectral shooting data from the „Resource-P" spacecraft No. 1. Two datasets such as AqwaObj and AntroObj have been created on the basis of these spectral standards.

The text file with the HSF (figure 1) contains information represented as the two-element tuples separated by a semicolon. In each tuple, the first number is the wavelength (nm) in the form (it is integer number), the second number is the value of the SRC corresponding to the wavelength, represented as the real number from the range [0; 1] with 4 bits of the fractional part, separated from the entire part by a dot. The data in the tuple is separated by a comma. The tuples are ordered by increasing of the wavelength values.

Firstly, all standard HSFs were combined into the common dataset of 220 objects with 127 wavelengths (the SRCs of all objects for these wavelengths were different from zero). The wavelengths were distributed in the range from 401 to 971 nm with non-uniform step. The AqwaObj dataset was created from this dataset by specifying the membership class for the uploaded HSF. The membership class of the real water object was designated as the "class 1", the class of any other object was designated as the "Class 2". Similarly, the AntroObj dataset was created.

The AqwaObj dataset is the sample of 220 HSFs, 53 of which belong to water objects (class 1), the remaining 167 HSFs belong to objects of different nature (class 2).

The AntroObj dataset is the sample of 220 HSFs, 80 of which belongs to anthropogenic objects (class 1), the remaining 140 HSFs belong to objects of different nature (class 2).

To develop the SVM classifier, it is necessary: to determine the value of the regularization parameter, which allows finding a compromise between the maximization of the strip separating the classes and minimizing of the overall error of the SVM classifier; to specify the kernel function; to determine the parameters values of the kernel function; to form the training and test sets.

The classification quality can be estimated using the following indicators: overall accuracy $Accur$; sensitivity Se; specificity Sp; F_1-measure; AUC, calculated at the test set (AUC_{test}); a number of errors of the I-st and II-nd type (Er_I and Er_{II}), number of errors at the training and test sets (Er_{train} and Er_{test}).

The kernels with polynomial, radial basis, sigmoid and linear kernel functions with the parameters values set by default were included in the search. The test set was randomly generated and contained 20% of objects from the initial dataset. The remaining 80% of objects from the initial dataset were included in the training set.

Table 1 shows the results of the SVM classifiers development. For each dataset the values of $Accur$ (overall accuracy), F_1-measure, the overall number of errors of the classifier that showed the best classification results (the highest values of $Accur$ and F_1-measure, the smallest number of errors at the training and test sets) are selected.

Table 1 shows that even at the kernel parameters values set by default the SVM algorithm allows to construct the classifier with the overall classification accuracy exceeding 90% for all datasets that demonstrates its high generalizing ability. Moreover, for the AqwaObj dataset the overall classification accuracy is equal to 98.64%, therefore, there is no need for increase in accuracy of classification and search of optimum parameters values for this dataset.

For the AntroObj dataset we developed 18 private SVM classifiers with the radial basis function rbf and the different parameters values of C and σ. The different training and test sets randomly generated from the original data set were used. The different training and test sets randomly generated from the original data set were used. The values of parameters and quality indicators of 18 private classifiers have been shown in the table 2. At the training for each private SVM classifier the training set was formed in a random way on the basis of the initial experimental dataset. The number of objects in the test set was equal to 10% – 30% of the initial number of objects in the initial experimental dataset.

Table 1. The results of the SVM classifier development with the values of the kernel function parameters set by default

Dataset $(s \times q)$ $(Train/Test)$	The kernel Function type	The quality assessment of the SVM classifier										
								The number of errors				
		$Accur$, %	F_1, %	Se, %	Sp, %	AUC_{test}, %	SV	Er_{train}	Er_{test}	Er_I	Er_{II}	Overall sum
AqwaObj	*lin*	95.91	91.59	92.45	97.01	0.9339	38	8	1	4	5	9
(220×127)	***poly***	**98.64**	**97.20**	98.11	98.80	0.9181	31	0	3	1	2	**3**
$(176/44)$	*rbf*	98.18	96.23	96.23	98.80	0.9256	148	0	4	2	2	4
	mlp	6.06	71.82	3.77	93.41	0.3281	31	48	14	51	11	62
AntroObj	*lin*	71.82	68.37	83.75	65.00	0.8047	143	53	9	13	49	62
(220×127)	***rbf***	**90.45**	**87.72**	93.75	88.57	0.7294	154	10	11	5	16	**21**
$(176/44)$	*mlp*	64.09	42.34	36.25	80.00	0.3945	80	62	17	51	28	79

At the testing it was found, that the private classifiers indicate the classification accuracy ranged from 84.09% to 92.73%, and the initial values of the determination coefficient, calculated for all 18 private classifiers, are in the range from 0.615 to 0.925. As a result, the threshold values θ were examined from the range [0.60; 0.95] with step 0.05. Values of the classification parameters corresponding to the different threshold values θ for the AntroObj dataset are given in the table 3.

Table 2. The values of parameters and quality indicators of the private classifiers (AntroObj dataset)

Classifier number	1	2	3	4	5	6	7	8	9	10	11	12	13	14	15	16	17	18
C	1	1	1	1	0.5	0.5	0.5	1.5	1.5	1.5	1.5	1.5	2	2	2	2	2	0.5
σ	1	1.5	2	2.5	1	1.5	2	0.5	1	1.5	2	2.5	0.5	1	1.5	2	2.5	0.5
$Accur$	89.09	87.73	87.73	84.09	87.27	85.91	85.00	91.82	92.27	89.09	87.27	86.36	90.45	92.73	90.00	89.09	88.18	90.45
SV	141	141	140	137	163	147	158	151	146	127	135	136	157	159	140	131	129	184
$Train$	154	176	187	198	176	165	198	154	165	165	187	198	165	187	176	187	198	187
Er_{train}	8	15	19	27	19	19	26	0	3	13	20	24	0	6	12	15	21	4
$Test$	66	44	33	22	44	55	22	66	55	55	33	22	55	33	44	33	22	33
Er_{test}	16	12	8	8	9	12	7	18	14	11	8	6	21	10	10	9	5	17
Se	86.25	86.25	90.00	91.25	97.50	88.75	93.75	83.75	90.00	91.25	91.25	92.50	76.25	92.50	93.75	88.75	91.25	98.75
Sp	90.71	88.57	86.43	80.00	81.43	84.29	80.00	96.43	93.57	87.86	85.00	82.86	98.57	92.86	87.86	89.29	86.43	85.71
Er_I	13	16	19	28	26	22	28	5	9	17	21	24	2	10	17	15	19	20
Er_{II}	11	11	8	7	2	9	5	13	8	7	7	6	19	6	5	9	7	1
AUC_{test}	0.804	0.779	0.819	0.719	0.883	0.853	0.641	0.785	0.775	0.789	0.698	0.752	0.766	0.708	0.870	0.808	0.925	0.615

θ_0	0.451	0.606	0.765	0.787	0.608	0.689	0.833	0.211	0.381	0.682	0.679	0.707	0.175	0.399	0.585	0.605	0.751	0.160

Table 3. The values of the classification quality indicators for the different thresholds of the determination coefficient (AntroObj dataset)

Classification quality indicator	Strategy	The threshold value of the determination coefficient					
		0.95-0.85	**0.8**	**075**	**0.7**	**0.65**	**0.6**
Accur, %	Majority vote	91.82	91.82	91.82	91.82	91.82	91.82
	Maximum and minimum	96.36	96.36	96.36	96.82	97.73	**98.64**
	Median and sum	92.27	92.27	94.09	95.45	96.82	**98.64**
	Product	92.27	92.27	94.09	95.45	96.82	**98.64**
Se, %	Majority vote	97.50	97.50	97.50	97.50	97.50	97.50
	Maximum and minimum	96.25	96.25	96.25	96.25	97.50	98.75
	Median and sum	95.00	95.00	97.50	97.50	98.75	100
	Product	95.00	95.00	97.50	97.50	98.75	100
Sp, %	Majority vote	88.57	88.57	88.57	88.57	88.57	88.57
	Maximum and minimum	96.43	96.43	96.43	97.14	97.86	98.57
	Median and sum	90.71	90.71	92.14	94.29	95.71	97.86
	Product	90.71	90.71	92.14	94.29	95.71	97.86
Er_I	Majority vote	16	16	16	16	16	16
	Maximum and minimum	5	5	5	4	3	2
	Median and sum	13	13	11	8	6	3
	Product	13	13	11	8	6	3
Er_{II}	Majority vote	2	2	2	2	2	2
	Maximum and minimum	3	3	3	3	2	1
	Median and sum	4	4	2	2	1	0
	Product	4	4	2	2	1	0
Number of classifiers in the ensemble		18	17	14	13	10	7

The optimal threshold value θ^* for the reviewed example equals to 0,6, since for the threshold value $\theta^* = 0.6$ all five classification rules (11) – (15) give the stable improvement of the classification quality when the number of classifiers reduces to the number corresponding to the threshold value $\theta^* = 0.6$. In this case, the finite number of classifiers in the SVM ensemble is equal to 7. A further decrease in the number of classifiers is not feasible (due to a further sharp decrease in their number and a substantial reduction of their variety).

The use of all five strategies with $\theta^* = 0.3$ allowed classifying correctly 98.64% of the objects of the initial dataset. At the same time, the maximum classification accuracy of one of the private SVM classifiers, used in the SVM ensemble, was equal to 92.73%, and the accuracy reached with the use of the majority vote rule was equal to 91.82%. Thus, the use of the SVM ensemble allowed increasing the classification accuracy almost by 6% compared to the maximum classification accuracy of one of the private classifiers in the SVM ensemble.

5. Conclusions

The SVM ensembles based on the decorrelation maximization algorithm for the different strategies of the decision-making on the data classification and the majority vote rule allow reducing the accident classification decision received by one classifier, and help to improve the classification accuracy. The shortcomings of some private classifiers are compensated by strengths of others private classifiers thanks to combination of their results. Classifiers counterbalance the results' accident of each other, finding the most plausible output classification decision. It allows finding the best classification result with minimum classification error.

The experimental results obtained on the basis of the hyperspectral shooting data confirm, that the SVM ensembles allow increasing the classification quality of objects by their GSFs. However, it is necessary to create the representative dataset of the HSFs and use it to learn the SVM classifiers.

References

[1] Demidova L A and Truhanov S V 2015 *Contemporary Engineering Sciences* **8** (20) 885.
[2] Grigorieva O, Brovkina O, Mochalov V, Akhtman Y, Zelentsov V, Potryasaev S, Kozyr I and Belova N 2016 *The 4th International Workshop on Simulation for Energy, Sustainable Development and Environment*.
[3] Taganov A I, Kolesenkov A I, Babaev S I and Sablina V A (2017) *6th Mediterranean Conference on Embedded Computing (MECO) Proceedings*. Budva, Montenegro pp. 130–133.
[4] Sablina V A, Novikov A I, Nikiforov M B and Loginov A A (2013) *2nd Mediterranean Conference on Embedded Computing (MECO) Proceedings*, Budva, Montenegro pp.117–120.
[5] Chapelle O, Vapnik V, Bousquet O and Mukherjee S 2002 *Machine Learning* **46** 131.
[6] Yu L, Wang S, Lai K K, and Zhou L 2008 *Springer* 244.
[7] Demidova L, Nikulchev E and Sokolova Yu 2016 *International Journal of Advanced Computer Science and Applications* **7**(5) 294.
[8] Graf H P, Cosatto E, Bottou L, Durdanovic I and Vapnik V 2005 *Advances in Neural Information Processing Systems* **17** 521.
[9] Eastaff M S and Premalatha P 2015 *Proceedings of the UGC Sponsored National Conference on Advanced Networking and Applications* pp. 191–193.

IOP Conf. Series: Materials Science and Engineering **312** (2018) 012006 doi:10.1088/1757-899X/312/1/012006

Adaptive neural network motion control for aircraft under uncertainty conditions

A V Efremov, M S Tiaglik and Yu V Tiumentsev

Flight Dynamics and Control Department, Moscow Aviation Institute (National Research University), Moscow, Russia

E-mail: yutium@gmail.com, pvl@mai.ru

Abstract. We need to provide motion control of modern and advanced aircraft under diverse uncertainty conditions. This problem can be solved by using adaptive control laws. We carry out an analysis of the capabilities of these laws for such adaptive systems as MRAC (Model Reference Adaptive Control) and MPC (Model Predictive Control). In the case of a nonlinear control object, the most efficient solution to the adaptive control problem is the use of neural network technologies. These technologies are suitable for the development of both a control object model and a control law for the object. The approximate nature of the ANN model was taken into account by introducing additional compensating feedback into the control system. The capabilities of adaptive control laws under uncertainty in the source data are considered. We also conduct simulations to assess the contribution of adaptivity to the behavior of the system.

1. Introduction

We need to provide motion control of modern and advanced aircraft under diverse uncertainty conditions [1, 2]. The aircraft control system should be able to adapt quickly to the current changes in the situation to ensure flight safety in such conditions. We analyze in this article the capabilities of adaptive control algorithms concerning aircraft motion for the two well-known types of such systems, namely, MRAC (Model Reference Adaptive Control) and MPC (Model Predictive Control) [3-5]. Both these schemes of adaptive control (Figure 1) require a model of the control object. As shown in [6-12], in the case of a nonlinear control object, the most effective solution to the problem of adaptive control is the use of neural network technologies. These technologies are suitable for the development of both a control object model and a control law for the object.

2. Ensuring robustness of adaptive control system

The studied schemes of adaptive control essentially use the ANN models of the control object as a source of information about the behavior of this object. Since, due to the approximate nature of the ANN model, the real values of the variables describing the motion of the object inevitably differ from those obtained as the outputs of the ANN model, an error appears that reduces the quality of control. One possible approach to compensate for this error is that the inaccuracy of the ANN model we can interpret as some disturbance on the system. This disturbance causes the deviation of the trajectory of the real object from the reference trajectory. We can reduce the impact of this effect by introducing a compensating loop into the control system.

IOP Conf. Series: Materials Science and Engineering **312** (2018) 012006 doi:10.1088/1757-899X/312/1/012006

In the schemes shown in Figure 1, a PD-compensator is used. It realizes, through additional feedback, a correction signal of the form $u_{add} = K_p e + K_d \dot{e}$, where e is the difference between the outputs of the control object and the ANN model of this object. The simulation results show that, despite the simplicity, the compensating loop reduces the tracking error (the difference between the outputs of the control object and the reference model) about an order of magnitude.

As a model of the control object, a neural network such as NARX (Nonlinear AutoRegession with eXternal inputs) is used in the MRAC and MPC schemes [6-8]. The same network we use in the MRAC scheme as the neural controller. In the MPC scheme, the predictive controller implements the parametric optimization algorithm.

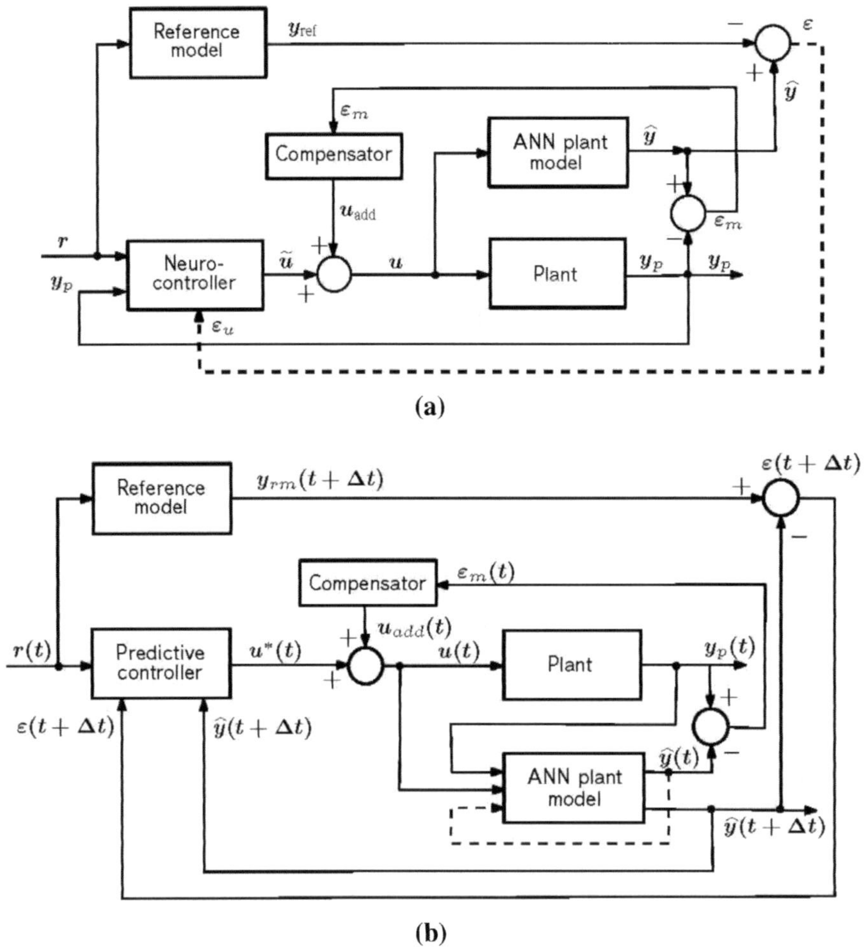

Figure 1. Adaptive control schemes: (a) Model Reference Adaptive Control (MRAC); (b) Model Predictive Control (MPC).

3. Adaptation to uncertainty in source data

With the adaptive control schemes shown in Figure 1, a series of simulations were carried out to evaluate the capabilities of such systems. As the control object, Hyper-X research hypersonic vehicle was used [13-16]. The uncertainty that we simulated in these computational experiments is an inaccurate knowledge of the dynamics of the control object. For such an imitation, the synthesis of the control law we apply to one flight mode (Mach number and flight altitude), and then the control law began to work under significantly different conditions.

The simulation results are shown in Figure 2 for the problem of stabilizing the longitudinal angular motion of Hyper-X vehicle. We use here the following notation: α is Hyper-X angle of attack; α_{ref} is reference signal for the angle of attack; e_{α} is the tracking error for the reference signal; φ_{act} is command signal for the elevator actuator. Here, we synthesize the control law for the $M = 7$, $H = 32$ km flight regime, and the testing was carried out for the $M = 5$, $H = 28$ km regime.

The results shown in Fig. 2, demonstrate the operation of the control system for two time intervals of 20 seconds each. On the first of them, the adaptation of the control law is carried out by feeding a perturbing signal on the input of the control system (this signal is an actively changing reference value α_{ref} for the angle of attack α). In the second interval with the same duration, the system is tested with a sequence of step input signals spaced apart in time so that the disturbed motion caused by the applied signal can attenuate until the next signal is applied. If this condition is met, then we can assume that the system in one session is tested by a set of independent step input signals, differing in magnitude.

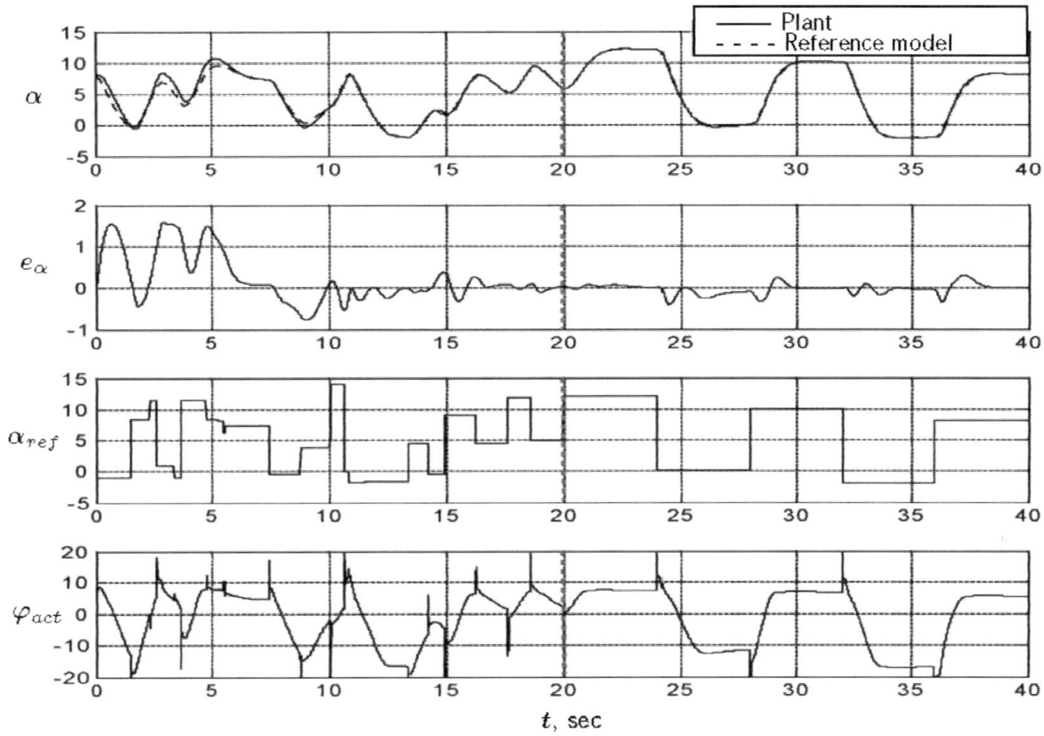

Figure 2. Adaptation to inexact source data for MRAC scheme (adaptation of the control law is for $0 \le t \le 20$ sec and testing for $20 \le t \le 40$ sec).

4. Efficiency evaluation of adaptation mechanisms

To evaluate the contribution of adaptivity and robustness in the system behavior, we conduct experiments in which we disconnect the adaptation mechanisms. The mechanism to ensure robustness of the system in the form of a compensating loop (Figure 1) remained switched on.

In the performed experiments, the results of which for MRAC scheme are presented in Figure 3, the operation time of the system was divided, as in the experiments with robustness (Figure 2), into two segments with a duration of 20 seconds each. The nature of the signal α_{ref} that is the reference value for the angle of attack is the same as in the previous case (Figure 2). Therefore, the first 20 seconds the reference signal is a random sequence of frequently changing values of the angle of attack. In this case, the perturbation from the input signal is not yet wholly damped, as the next reference signal acts on the object.

IOP Conf. Series: Materials Science and Engineering **312** (2018) 012006 doi:10.1088/1757-899X/312/1/012006

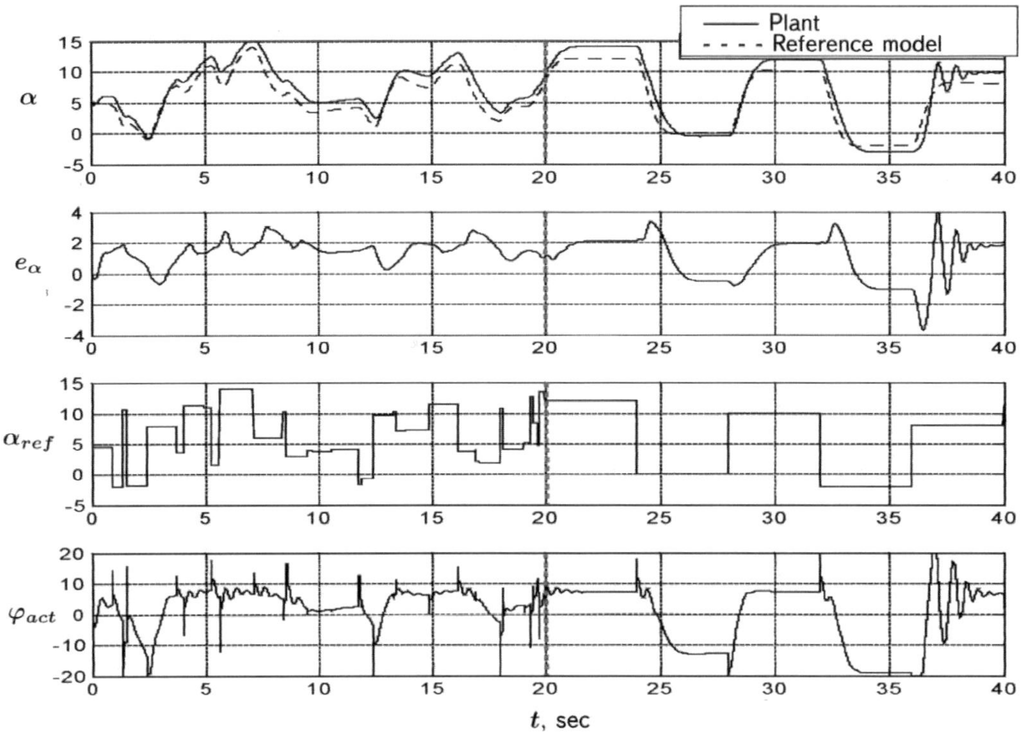

(a)

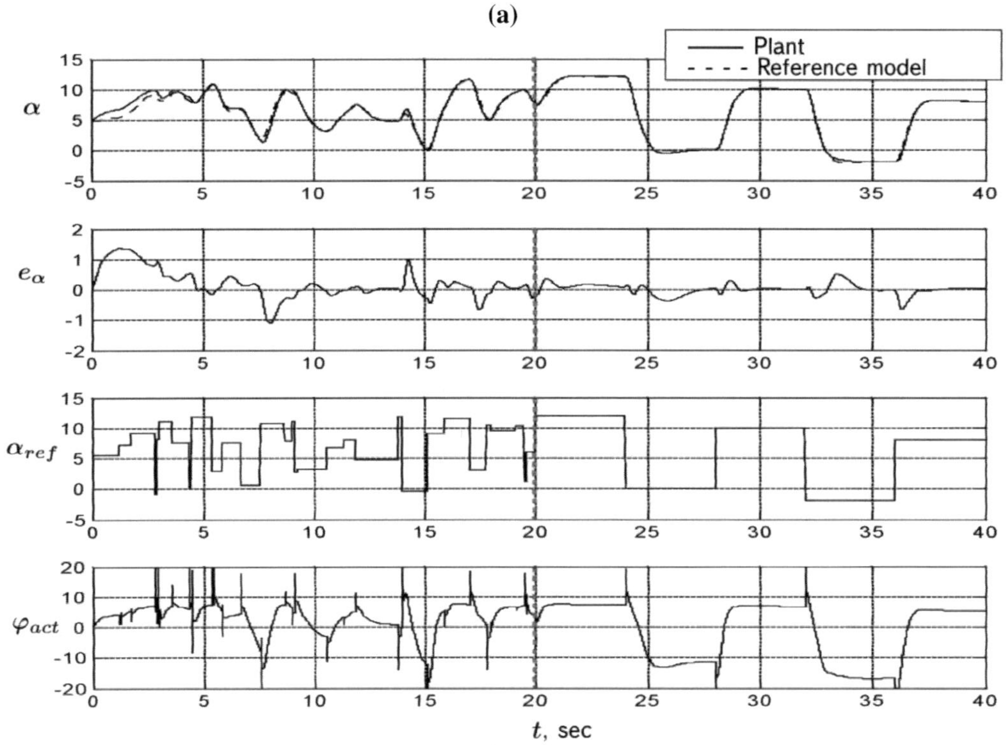

(b)

Figure 3. Efficiency evaluation of adaptation mechanisms for MRAC scheme: (a) the adaptation mechanism is disabled; (b) the adaptation mechanism is enabled (adaptation is for $0 \le t \le 20$ sec and testing for $20 \le t \le 40$ sec).

This kind of complicated reference signal allows us to estimate the cumulative effect of several disturbances, the reactions from the impact of which overlap in time. This approach makes it possible to test the control system under rather severe conditions. In the second time interval, the control system is tested using a more traditional input signal, which makes it possible to evaluate the behavior of the system for isolated step disturbances. As we can see from the results presented in Figure 3, the lack of adaptation mechanisms leads to a sharp deterioration in the quality of motion control. Namely, values of the tracking error $e_\alpha = \alpha_{ref} - \alpha$ are within $\pm 2^0$, and, in some cases, even within $\pm 4^0$. For the MPC system, the results are similar.

Thus, the adaptation mechanisms significantly reduce the tracking error and expand the frequency band in which the error does not exceed some preset value. It follows that if a system has adaptive properties, this allows it to deal with a much broader class of parametric uncertainties in control objects.

5. Conclusion

The general conclusion, based on the results presented in this article, is that the methods of nonlinear adaptive-robust modeling and control in MRAC and MPC variants are a powerful and promising tool that allows us to efficiently solve the movement control tasks for aircraft under uncertainty conditions.

Acknowledgments

The results included in the paper are sponsored by the Ministry of Education and Science (contract No. 9.7465.2017/БЧ)

References

[1] Xiao L and Gao A 2015 *Vehicle System Dynamics: Int J Vehicle Mechanics and Mobility* **48**(10) 1167–92
[2] Piche S, Sayyar-Rodsari B, Johnson B and Gerules M 2000 *IEEE Control Systems Magazine* **20** (3) 53–62
[3] Astolfi A, Karagiannis D and Ortega R 2008 *Nonlinear and adaptive control with applications* (Berlin: Springer)
[4] Ioannou PA and Sun J 1995 *Robust adaptive control* (Upper Saddle River, NJ: Prentice Hall)
[5] Tao G 2014 Multivariable adaptive control *Automatica* **50** (11) 2737–64
[6] Hagan M T, Demuth H B, Beale M H and De Jesus O 2014 *Neural network design*, 2nd Ed (Longview, TX: PSW Publishing Co)
[7] Dreyfus G 2005 *Neural networks: Methodology and applications* (Berlin: Springer)
[8] Mandic D P and Chambers J A 2001 *Recurrent neural networks for prediction: Learning algorithms, architectures and stability* (New York: John Wiley & Sons)
[9] De Jesus O and Hagan M T 2007 *IEEE Trans on Neural Networks* **18** (1) 14–27
[10] Omidvar O M and Elliott D I (Eds.) 1997 *Neural systems for control* (Cambridge, Mass.: Academic Press)
[11] Soloway D and Haley P J 1997 *NASA TM 110244* 18 pp
[12] Kondrat'ev A I and Tyumentsev Yu V 2013 *Russian Aeronautics (Iz.VUZ),* **56** (2) 135–144, (3) 257–265
[13] Davidson J, Lallman J D et al 1999 *AIAA-99-4124* 11 pp
[14] Engelund W C, Holland S D et al 2000 *AIAA-2000-4006* 14 pp
[15] Holland S D, Woods W C and Engelund W C 2001 *Journal of Spacecraft and Rockets* **38** (6) 828–835
[16] Morelli E A, Derry S D and Smith M S 2005 *AIAA-2005-5921* **15**

IOP Conf. Series: Materials Science and Engineering **312** (2018) 012007 doi:10.1088/1757-899X/312/1/012007

Predictive display design for the vehicles with time delay in dynamic response

A V Efremov, M S Tiaglik, I H Irgaleev and E V Efremov

Flight Dynamics and Control Department, Moscow Aviation Institute (National Research University), Moscow, 125993, Russia

E-mail: pvl@mai.ru

Abstract. The two ways for the improvement of flying qualities are considered: the predictive display (PD) and the predictive display integrated with the flight control system (FCS).
The both ways allow to transforming the controlled element dynamics in the crossover frequency range, to improve the accuracy of tracking and to suppress the effect of time delay in the vehicle response too. The technique for optimization of the predictive law is applied to the landing task. The results of the mathematical modeling and experimental investigations carried out for this task are considered in the paper.

1. Introduction

The precise path control task is a one of the most difficult manual control task as for the aeronautical and space vehicles because of their controlled element dynamics is characterized by the high order pole at the origin. For example the height-to-inceptor transfer function of aircraft in the landing task as well as the spacecraft in the docking task with International Space Station (ISS), has the second order pole at the origin [1, 2]. It requires the considerable pilot lead compensation what decreases the accuracy [3, 4]. Except this peculiarity, the highly augmented aircraft dynamics is characterized by the time delay [5, 6]. Such delay is caused by different filters, actuator dynamics, digital realization of aircraft flight control system laws. It achieves 0,1÷0,2 sec and even higher. It requires the considerable pilot lead compensation to avoid the pilot-induced oscillations in the closed-loop system [7]. The teleoperator control used for the docking of spacecraft with ISS and UAV (unmanned aerial vehicle) control implemented via satellite is accompanied by higher time delay (up to 1 sec for spacecraft [2], and 0.5 sec for UAV control [8]). The manual control is much more complicated in these cases. Because of it the automatic control is used more and more often.

The problem discussed in the paper is dedicated to the compensation of the time delay in their response by use of the predictive display.

2. General technique for the predictive display design.

The potentiality effect of predictive display for the landing of aircraft on the runway was shown in [9]. Such type of display was offered in [10] for the tasks of refuelling and landing the carrier. It displays 3D corridor within which the aircraft has to move. Such information allows to evaluate the current and future position of the aircraft and to transform pilot behaviour from compensatory to preview type. The predictive display shows also the surface moving inside the corridor with the aircraft velocity V, where the predictive angle $\varepsilon_{pr} = \gamma_{pr} + \frac{\Delta H}{L_{pr}}$ is projected. Here $L_{pr} = T_{pr} \cdot V$ is the distance between a pilot and the moving surface, ΔH is a deviation of aircraft altitude. The equation for predictive angle γ_{pr} depends on the controlled element dynamics. It is shown in [10] that for a conventional aircraft with the practical zero time delay in responses the predictive angle is defined by the equation $\gamma_{pr} = \gamma + \frac{T_{pr}}{2}\dot{\gamma}$. The use of such law allows to transform the controlled element dynamics providing the slope of its amplitude frequency response characteristics ($W_c^* = \frac{\varepsilon_{pr}(j\omega)}{x_e(j\omega)}$) in the crossover frequency range close to -20 dB/dec. It increases the accuracy and decreases the pilot workload. The technique for the definition of T_{pr} is considered in [10, 11]. It is based on the minimization variance of error ΔH ($\sigma^2_{\Delta H}$) (see the block diagram in fig.1). The procedure of minimization supposes the use of the modified pilot structural model considered in [10, 11].

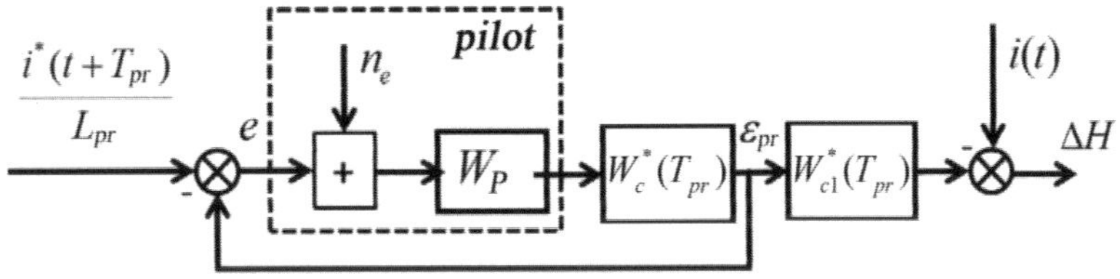

Figure 1. Pilot-aircraft system with predictive display

The results of preliminary selected T_{pr} are verified then in the ground simulation. The developed technique was adopted for investigation of the following manual control tasks:

1. Predictive display law synthesis for compensation of time delay in:
- aircraft landing;
- docking of spacecraft with the ISS.
2. The integration of predictive display and flight control system.

The analysis of the different alternative laws demonstrated that the general equation for predictive angle for all considered tasks has the following form.

$$\varepsilon_{pr} = \frac{\Delta H(t-\tau)}{L_{pr}} + \gamma_M(t) + \dot{\gamma}_M(t)\frac{T_{pr}}{2} + [\gamma(t-\tau) - \gamma_M(t)]W_f, \quad (1)$$

where $W_f = \frac{K}{T_f p + 1}$, $\gamma_M(t)$ and $\dot{\gamma}_M(t)$ are calculated (with help of on-board computer) path angle and derivative, τ - time delay. The parameters T_f and K are selected by minimization of variance $\sigma^2_{\Delta H}$.

3. Predictive display law synthesis for the time delay compensation.

The principle of time delay compensation is illustrated by the fig. 2.

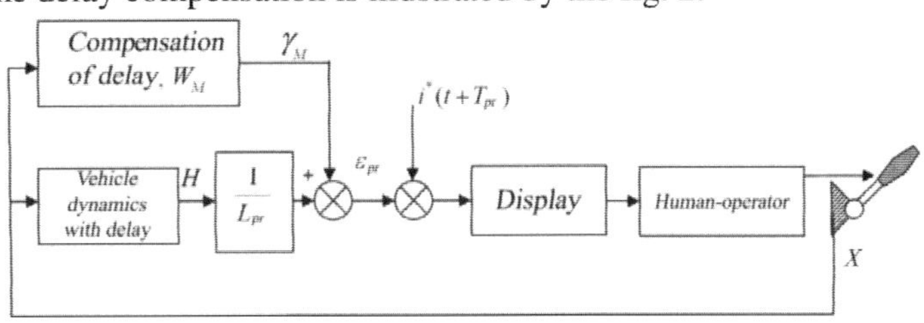

Figure 2. Compensation of time delay

Here $i^*(t+T_{pr})$ - the predictive program trajectory projected on the display screen in the form of 3D corridor with the centre line. The predictive angle ε_{pr} is projected on the surface moving inside the 3D corridor.

In the landing task of the highly augmented aircraft the nonzero time delay in its dynamic can be suppressed by the law

$$\varepsilon_{pr} = \frac{\Delta H(t-\tau)}{L} + \gamma_m(t) + \dot{\gamma}_m(t)\frac{T_{pr}}{2}, \quad (2)$$

which is a special case of equation (1), derived from condition $W_f = 0$. The mathematical modelling of pilot-aircraft system allowed to get optimum value T_{pr}. In particular it was obtained that $T_{pr_{opt}} = 1.4$ sec what is higher than T_{pr} obtained for the case when the time delay is zero.

It can be shown that for the spacecraft teleoperator control height motion in the docking task the summand $\dot{\gamma}_m(t)\frac{T_{pr}}{2}$ and W_f in eq. (1) have to be zero. In that case this equation transforms in the following

$$\varepsilon_{pr} = \frac{\Delta H}{L_{pr}} e^{-\tau s} + \gamma_M, \quad (3)$$

It provides the following controlled element dynamics transfer function $W_C = \frac{K(T_{pr}s + e^{-\tau s})}{T_{pr}s^2(Ts+1)}$ (T – time constant of the thruster dynamics). The use of predictive information provides the slope of $\frac{dlg|W_C|}{dw} \cong -20\,dB/dec$ in the range 0.6÷6 1/sec what improves the accuracy considerably. The optimization of T_{pr} allowed to get its optimum value equal to 16÷18 sec.

The experiments on ground-based simulators demonstrated that the use of predictive display allows to suppress the time delay effect considerably (see fig. 3, where $\sigma^2_{\Delta H}$ - variance of error in glideslope tracking, $\sigma^2_{\Delta H \max}$ - variance of error in glideslope tracking for the case of uncompensated time delay).

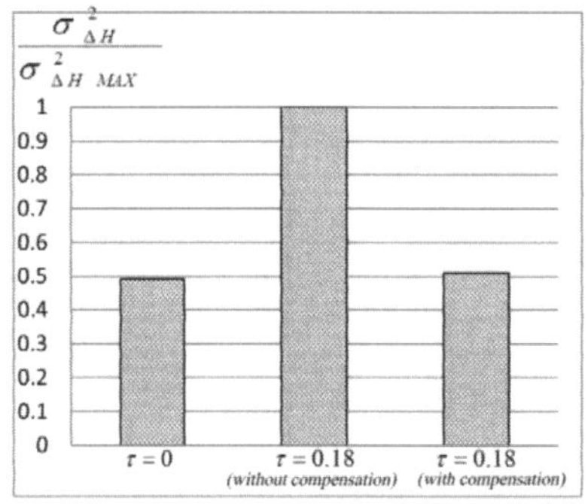

Figure 3. The results of ground-based simulation

The use of predictive display allows decreasing the docking error by about 3 times.

4. The integration of predictive display and the flight control system.

In the frame of the considered investigation the integration of predictive display with the flight control system of the aircraft employing the direct lift control (DLC) was studied. This surface was used for provision of control of the angle of attack changes without having to change the pitch angle (α = var, $\vartheta = const$) [12, 13]. The investigation demonstrated that realization of such mode requires the installing a prefilter into the flight control system $W_{pr} = \frac{s+\omega_f}{s}$ (where $\omega_f = 0.2\ 1/\sec$) for suppression the static error in glide slope tracking. The integration of flight control system with predictive display required to modify the equation (1) for predictive angle and to eliminate the summand $\dot{\gamma}\frac{T_{pr}}{2}$ in it. Optimization of predictive display law demonstrated that the increase of time delay required the proportional increase of predictive time T_{pr} too. The experimental investigations executed on ground-based simulator confirmed that the integration of DLC and display allows to suppress the effect of time delay in glideslope tracking task in turbulence and calm atmosphere conditions (fig. 4, 5) and to improve the accuracy in several times.

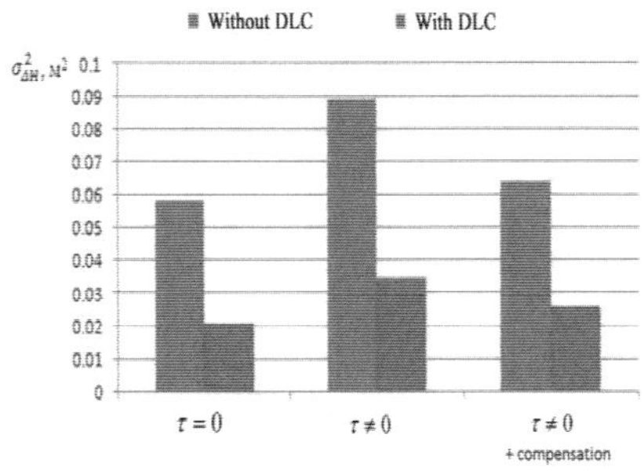

Figure 4. Effect of DLC use (mathematical modeling)

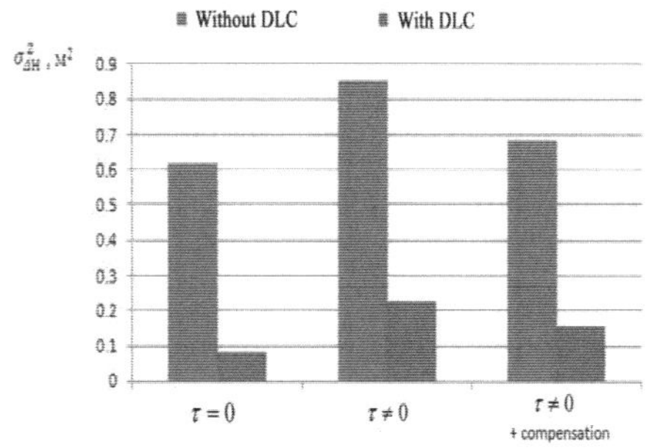

Figure 5. Effect of DLC use (experimental investigations)

Acknowledgments
The results included in the paper are the results of the investigation were sponsored by the Ministry of Education and Science (contract № 9.7465.2017/BP).

References

[1] D. McRuer, J. Ashkenas, D. Graham 1973 *Aircraft dynamics and automatic control* (Princeton, New Jersey : Princeton University)
[2] Efremov A.V., Tjaglik M.S., Tiumentzev U.V., Wenqian T. 2016 Pilot behavior modeling and its application to manual control tasks *IFAC-PapersOnLine* **49** (32) 159-164
[3] D. McRuer, E. Krendel 1974 *Mathematical models of pilot behavior* AGARD-AG-188
[4] D. McRuer et al 1965 *Human pilot dynamics in compensatory systems* AFFDL-TR-65-15
[5] D. McRuer 1998 *Aviation safety and pilot control: understanding and preventing unfavourable pilot-vehicle interactions* (Washington: National Academy of Science)
[6] D. Klyde, D. McRuer, T. Myers 1995 *Unified pilot-induced oscillation theory Vol. 1 Analysis with linear and nonlinear effective vehicle characteristics* WL-TR-96-3028
[7] Efremov A.V., et al 1992 *Pilot as a dynamic system* (Moscow:Mashinostroenie) (in Russian)
[8] Feng Wang, Sheng Jun Qi, L.Jing 2017 An analysis of time delay for remote pilot vehicle *MATEA Web of Conference* **114** 04012
[9] G. Sachs 2000 Perspective Predictor/Flight – Path Display with Minimum Pilot Compensation *Journal of Guidance, Control, and Dynamics* **23**(3) 51-64
[10] Efremov A.V., Tjaglik M.S. 2011 The Development of Perspective Displays for Highly Precise Tracking Tasks *Advances in Aerospace Guidance, Navigation and Control* (New-York: Springer)
[11] Efremov A.V., Tyaglik M.S. 2010 Development of displays provided high accuracy control *Journal Flight* 11 9-19, (in Russian)
[12] M.J. Pinsker 1970 The control characteristics of aircraft employing direct – lift control US *Ministry of technology report R. and M. № 3629*
[13] I. Guskov, G. Zagaynov 1980 *Flight control* (Moscow:Mashinostroenie) (in Russian)

IOP Conf. Series: Materials Science and Engineering **312** (2018) 012008 doi:10.1088/1757-899X/312/1/012008

Neural network identification of aircraft nonlinear aerodynamic characteristics

M V Egorchev and Yu V Tiumentsev

Flight Dynamics and Control Department, Moscow Aviation Institute (National Research University), 125993, Volokolamskoe shosse, 4, Moscow, Russia

E-mail: mihail.egorchev@gmail.com, yutium@gmail.com

Abstract. The simulation problem for the controlled aircraft motion is considered in the case of imperfect knowledge of the modeling object and its operating conditions. The work aims to develop a class of modular semi-empirical dynamic models that combine the capabilities of theoretical and neural network modeling. We consider the use of semi-empirical neural network models for solving the problem of identifying aerodynamic characteristics of an aircraft. We also discuss the formation problem for a representative set of data characterizing the behavior of a simulated dynamic system, which is one of the critical tasks in the synthesis of ANN-models. The effectiveness of the proposed approach is demonstrated using a simulation example of the aircraft angular motion and identifying the corresponding coefficients of aerodynamic forces and moments.

1. Introduction

We proposed in [1-3] a semi-empirical approach to ANN-modeling of nonlinear controlled dynamical systems. This approach is based on combining theoretical knowledge about the object of modeling and experimental data on its behavior. The result is a model of the gray box type, which can be configured by machine learning methods. In the problems of modeling the motion of aircraft, the traditional theoretical model, widely used in the solution of various applied tasks, is a system of ordinary differential equations (ODE). This model, in combination with the corresponding experimental data, can be used to form a semi-empirical neural network model of aircraft motion. The source theoretical model of the aircraft motion model [4] contains, as constituent elements, expressions for the dimensionless coefficients of aerodynamic forces and moments. One of the variants of the classical problem of system identification [5-8] is related to them, namely, the problem of identifying aerodynamic characteristics of aircraft [9-15], which consists in restoring the relationships for the nondimensional coefficients of forces and moments from the available experimental data on the behavior of the aircraft. In the problem considered below, as the source theoretical model, we use a system of ordinary differential equations describing the total angular motion of the aircraft. The semi-empirical neural network model formed on this basis includes five black-box ANN-modules, describing the coefficients of the lift and lateral forces, the pitch, yaw, and roll moment coefficients, each of which depends nonlinearly on several parameters of the aircraft motion. These five relationships need to be found (recovered) basing on available experimental data for the observed variables of the dynamic system, i.e., we have to solve the problem of identifying the aircraft aerodynamic characteristics. We confirm the effectiveness of the proposed approach using the corresponding simulation results.

2. Comparison of the proposed and traditional approaches to the identification of aerodynamic characteristics of an aircraft

The proposed approach to the identification of aerodynamic characteristics of an aircraft differs substantially from the traditionally accepted way to solving problems of this class. Namely, with the traditional approach [9-15], as a rule, we use a linearized model of the disturbed motion of an aircraft. In this case, the relationships for the aerodynamic forces and moments, acting on the aircraft, are represented in the form of expanding them into a Taylor series, leaving only first-order terms in them (in rare cases also second-order terms). Then the solution of the identification problem is reduced to obtaining from the experimental data the relations describing the coefficients of the Taylor expansion. The values of these factors are determined, first of all, by the value of the partial derivatives of the dimensionless coefficients of the aerodynamic forces and moments by the various parameters of the aircraft motion (C_{L_α} , C_{m_α} , C_{l_β} , C_{m_q} etc.).

In contrast, the semi-empirical approach realizes the derivation of the relationships for the coefficients of the forces C_D, C_L, C_z and the moments C_l, C_n, C_m as holistic nonlinear dependencies from the corresponding arguments, without using their Taylor expansions. Thus, the proposed approach evaluates the functions themselves, and not the coefficients of their Taylor expansions. We implement each of these dependencies as a separate ANN-module, built into the semi-empirical model of the aircraft motion.

Sometimes the derivatives C_{L_α} , C_{m_α} , C_{l_β} , C_{m_q} etc. are necessary for solving some problems, for example, for analysis of the stability and controllability characteristics of an aircraft. In this case, it is easy to find them using the appropriate ANN-modules obtained in the formation of the semi-empirical ANN-model of the aircraft motion.

3. Semi-empirical ANN-model of aircraft three-axis angular motion

To estimate the efficiency of the proposed approach, let us consider the problem of simulating the aircraft three-axis rotational motion. We describe it by a typical set of equations for aircraft flight dynamics [4]. This set consists of fourteen differential equations, which we do not present here because of it being too complicated. The state variables of the dynamic system are the roll angular rate p, the pitch angular rate q and the yaw angular rate r (degree/second); the roll angle Φ, the yaw angle Ψ and the pitch angle Θ (degree); the angle of attack α, the angle of sideslip β; the angle of the all-moving tailplane deflection δ_e, the angle of the rudder deflection δ_r, the angle of the aileron deflection δ_a (degree); the angular rates of the all-moving tailplane, the rudder and the aileron deflections $\dot{\delta}_e$, $\dot{\delta}_r$, $\dot{\delta}_a$ (degree/second), respectively. The control inputs are the quantities δ_e^{act} , δ_r^{act} , δ_a^{act} that are command driving signals supplied to the all-moving tailplane, the rudder and the aileron (degrees), respectively.

The given theoretical model contains six unknown nonlinear functions of several variables. These functions describe the dependences on the state variables for the coefficients of the axial $C_x(\alpha, \beta, \delta_e, q)$, the transverse $C_y(\alpha, \beta, \delta_r, \delta_a, p, r)$ and the normal $C_z(\alpha, \beta, \delta_e, q)$ aerodynamic forces, respectively. They also define aerodynamic moments $C_l(\alpha, \beta, \delta_e, \delta_r, \delta_a, p, r)$, $C_m(\alpha, \beta, \delta_e, q)$ and $C_n(\alpha, \beta, \delta_e, \delta_r, \delta_a, p, r)$ of the roll, the pitch and the yaw, respectively. To implement these functions we include in the developing model six neural network modules in the form of the sigmoid feedforward neural networks with one hidden layer. The hidden layers include 1, 5, 3, 5, 10 and 5 neurons for the modules C_x, C_y, C_z, C_l, C_m and C_n, respectively.

Note, since in the model there are no controls acting on the acceleration/braking along the longitudinal axis of the aircraft, it is not possible to obtain the training set for the neural network module representing the drag coefficient C_x. This is the reason why we form the neural network module for C_x independently basing on the data from [20]. We embed it into the generated semi-empirical model. Then we "freeze" this module, that is, we disable variations of its adjustable parameters.

4. Generation of representative training set

When synthesizing the neural network models one of the critically important problems is the formation of a representative data set characterizing the behavior of the simulated dynamic system. Unfortunately, this problem has no simple solution, but it is crucial for obtaining an efficient and reliable model of the dynamic system.

In paper [3] we showed that for the class of problems we consider, the most effective way is to use the polyharmonic (multisine) actuating signal. According to this approach we form for each of m aircraft controls the input actions as the sum of harmonic signals each of which has its own phase shift φ_k. The input signal u_j corresponding to the j-th control surface has the form:

$$u_j(t) = \sum_{k \in I_j} A_k \cos\left(\frac{2\pi kt}{T} + \varphi_k\right), \quad j = 1, \ldots, m, \quad I_j \subset K, \quad K = \{1, 2, \ldots, M\},$$

$$u_j^*(t) = \tilde{u}_j(t) + u_j(t),$$

where M is the total number of harmonically connected frequencies, T is the interval of time during which the test actuating signal acts upon the dynamic system; A_k is the k-th amplitude of the sinusoidal component; $u_j^*(t)$ is the complete controlling action for the j-th control surface; $\tilde{u}_j(t)$ is the controlling action for the j-th control surface realizing the test maneuver.

When generating the training set as well as when testing the obtained semi-empirical neural network model the controlling actions affected the aircraft over all three channels simultaneously and at that the signals δ_e^{act}, δ_r^{act}, δ_a^{act} were formed as the polyharmonic ones when obtaining the training set and as random ones when testing the trained model.

5. Computer simulations

In our computer simulations using the theoretical model, the interval of time was $t \in [0, 20]$ seconds when obtaining data for the ANN-model learning, and it was $t \in [0, 40]$ seconds when testing the obtained neural network model. In both cases simulations were performed with the time step of Δt = 0.02 sec for the partially observable state vector $y(t) = [\alpha(t); \beta(t); p(t); q(t); r(t)]T$. The additive white noise with the standard deviations $\sigma\alpha = \sigma\beta = 0.02$ degree and $\sigma p = 0.1$ degree/sec, $\sigma q = \sigma r = 0.05$ degree/sec affected the system output y(t). If the neural network model reproduces the original system perfectly well, the noise affecting the system output defines the error of simulation completely. Consequently, comparison of the error of simulation with the standard deviation of the noise allows us to assess how well our simulations were. The standard deviation of the noise can be regarded as the target error of the simulations.

The learning was carried out basing on the sample (training set) $\{y_i\}$, $i = 1,\ldots,N$, obtained for the source theoretical model with the aid of the Matlab system using the Levenberg–Marquardt algorithm and the root-mean-square error (RMSE) criterion. We calculated the Jacobian matrix according to the RTRL (Real-Time Recurrent Learning) algorithm [16-19].

From Figure 1 we see that the errors for all the observable state variables are sufficiently small. Moreover, these errors do not increase with time, and this is the evidence of good generalization ability of the obtained neural network model. It must be emphasized that to ensure a great variety of the modeled system states as well as to ensure the largest possible variety of the differences between states at the adjacent instants of time, when testing the model we simulated very active work of the aircraft controls. An additional complicating factor was in providing the subsequent input disturbance affected the aircraft before the transition processes from one, or several preceding disturbances were finished.

IOP Conf. Series: Materials Science and Engineering **312** (2018) 012008 doi:10.1088/1757-899X/312/1/012008

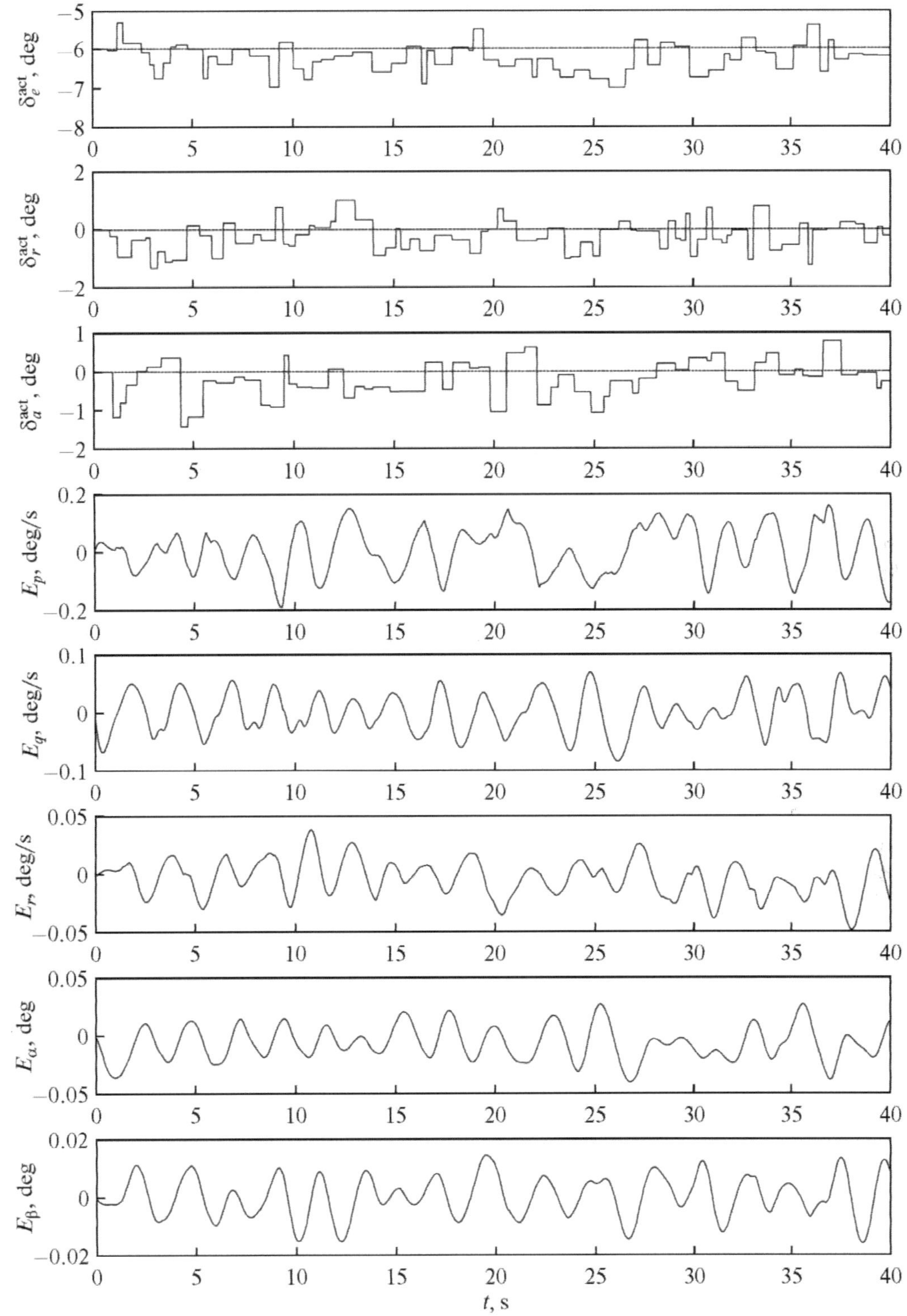

Figure 1. Estimation of generalizing properties for the neural network model after completing of the 1000-step of the learning process: E_α, E_β, E_p, E_q, E_r are the measures of inaccuracy of the corresponding observed data; the horizontal lines show the values of the controls corresponding to the test maneuver (the absolute altitude is 300 m, the airspeed is 148 m/sec)

We are also interested in the accuracy of solution when identifying the aerodynamic characteristics. We can estimate it comparing the values generated by the relevant neural network modules with experimental data [20]. The values of the root mean-square errors (RMSE) of the reconstruction of each function provided by the corresponding neural network modules are $\mathrm{RMSE}_{C_y} = 5.4257 \cdot 10^{-4}$, $\mathrm{RMSE}_{C_z} = 9.2759 \cdot 10^{-4}$, $\mathrm{RMSE}_{C_l} = 2.1496 \cdot 10^{-5}$, $\mathrm{RMSE}_{C_m} = 1.4952 \cdot 10^{-4}$, $\mathrm{RMSE}_{C_n} = 1.3873 \cdot 10^{-5}$. At that, the error level does not change significantly with time. We did not find the variations of the error that can influence negatively the adequacy of our semi-empirical neural network model.

6. Conclusion

The obtained results show that the methods of neural network modeling combined with knowledge and experience from the relevant subject area, as well as a representative training set, are a powerful tool for solving complicated problems for controlled dynamical systems of various classes. They also allow efficiently solving the problem of identifying aerodynamic characteristics of an aircraft.

Acknowledgments

This research is supported by the Ministry of Education and Science under Contract No. 9.9124.2017/ВУ.

References

[1] Egorchev M V, Kozlov D S, Tiumentsev Yu V and Chernyshev A V 2013 *J Computer and Inform Technology* No 9 pp 3–10 (In Russian)
[2] Egorchev M V, Kozlov D S and Tiumentsev Yu V 2014 *ICAS Paper 2014-0530* pp 1–8
[3] Egorchev M V and Tiumentsev Yu V 2015 *Optical Memory and Neural Networks (Information Optics)* **24** (3) pp 210–17
[4] Bochkariov A F et al 1985 *Aeromechanics of airplane: Flight dynamics* 2nd Ed. (Moscow: Mashinostroyeniye) (In Russian)
[5] Nelles O 2001 *Nonlinear system identification: From classical approaches to neural networks and fuzzy models* (Berlin: Springer)
[6] Sjöberg J et al 1995 *Automatica* **31** (12) pp 1691–1724
[7] Rubio J J and Yu W 2007 *Neurocomputing* **70** (13-15) pp 2460–6
[8] Singh M et al 2007 *IETE J of Research* **53** (1) pp 43–50
[9] Jategaonkar R V 2006 *Flight vehicle system identification: A time domain methodology* (Reston, VA: AIAA, Inc.)
[10] Klein V and Morelli E A 2006 *Aircraft system identification: Theory and practice* (Reston, VA: AIAA, Inc.)
[11] Hamel P G and Jategaonkar R V 1996 *J of Aircraft* **33** (1) pp 9–28
[12] Jategaonkar R V, Fischenberg D and von Gruenhagen W 2004 *J of Aircraft* **41** (4) pp 681–91
[13] Klein V 1989 *Progress in Aerospace Sciences* **26** (1) pp 1–77
[14] Morelli E A 2003 *13th IFAC Conf. on System Identification*, Paper REG-360 pp 1–7
[15] Juditsky A et al 1995 *Automatica* **31** (12) pp 1725–50
[16] Haykin S 2006 *Neural networks: A comprehensive foundation*: 2nd Ed (Amsterdam: Prentice Hall)
[17] Hagan M T, Demuth H B, Beale M H and De Jesus O 2014 *Neural network design*, 2nd Ed (Longview, TX: PSW Publishing Co)
[18] Dreyfus G 2005 *Neural networks: Methodology and applications* (Berlin: Springer)
[19] Mandic D P and Chambers J A 2001 *Recurrent neural networks for prediction: Learning algorithms, architectures and stability* (New York: John Wiley & Sons)
[20] Nguyen L T et al 1979 *NASA TP-1538* pp 1–223

IOP Conf. Series: Materials Science and Engineering **312** (2018) 012009 doi:10.1088/1757-899X/312/1/012009

Statics and buckling problems of aircraft structurally-anisotropic composite panels with the influence of production technology

L M Gavva and A I Endogur

Aircraft Design and Control Department, Moscow Aviation Institute (National Research University), Moscow, 125993, Russia

E-mail: rva101@mail.ru

Abstract. The mathematical model relations for stress-strain state and for buckling investigation of structurally-anisotropic panels made of composite materials are presented. The mathematical model of stiffening rib being torsioned under one-side contact with the skin is refined. One takes into account the influence of panel production technology: residual thermal stresses and reinforcing fibers preliminary tension. The resolved eight order equation and natural boundary conditions are obtained with variation Lagrange procedure. Exact analytical solutions for edge problems are considered. Computer program package is developed using operating MATLAB environment. The influence of the structure parameters on the level of stresses, displacements, of critical buckling forces for bending and for torsion modes has analyzed.

1. Introduction

The paper discusses the stress-strain state and buckling problems of flat rectangular multiplied panel made of composite materials, the casing of which an eccentrically supported by the longitudinal-transverse stiffening set. The panel is subjected to arbitrarily distributed transverse loading in the stationary temperature field and to distributed loading applied to the edges in casing plane. Boundary conditions at the contour are assumed to be of quite general type.

One should take into consideration the technological factors occurring in the manufacture of composites: residual thermal stresses arising during cooling after hardening and pre-stressed tension in reinforcing fibers.

The theory of thin-walled elastic robs is used without the hypothesis of the zero shear deformation for reinforcing elements being in the complex resistance – two-plane bending and limited torsion. Further development of the theory of thin-walled elastic robs related to the contact problem for the skin and the rib with improved rib model reflects the scientific novelty of the research.

The schematization of the panel as structurally anisotropic has been proposed as a design model when stress-strain state and critical forces of total bending form of buckling were determined. For many-waved torsion buckling study one should use the generalized functions set.

The problem of determining stress-strain state of structurally anisotropic panels is reduced to the solution of the boundary value problem for equation of the eighth order in the partial derivatives in a rectangular field. The differential equations of eighth order in partial derivatives are resolving for the pre-critical stress-strained state and for the buckling problem. The solution in closed form is designed by unitary trigonometric series for the particular case of conformable boundary conditions on two opposite sides and by the method of uniform solutions for arbitrary non-conformable boundary conditions at the contour. We examine all possible variants of the boundary edges restrictions in relation to the connecting plane problem and the bending one.

Computer program package in MATLAB has been performed for multi-criterion optimization for the design of structurally-anisotropic composite panels of FA.

2. Main relations of mathematical model

The papers [1–15] describe statics problems of structurally-anisotropic composite panels but the mathematical model offered lower is seemed to be new one.

The geometrical and physical relations are assumed at each *k*-ply on the macro-level for the skin accordingly to Kirchgoff hypothesis.

Stress-strain state components of each *k*-ply of composite stringers are calculated taking into account plate and rib displacement and rotation angle equalities. The cross-section warping is put free here. Shear strain occurring because of the thin-walled bar torsion is connected with its rotation relative to chosen pole and also relative to bending centre in view of «pure» torsion theory additives [16].

The problem consists in finding the base surface displacements.

Further discussion concerns the- mathematical model with the proposal of the infinitesimal of normal stresses caused by bar bending in panel plane and bar cross-section warping. We'll consider ribs working in tension-compression, in out- of-plane plate bending and in torsion.

Composite panel equilibrium equations and natural boundary conditions are formed by the variation method using Lagrange concept as a result of the minimization of the whole potential energy of the system.

Equilibrium differential equation system is written in operator form with linear differential operators. By using symbol integration method, when, for example, the first and the second expressions are identically satisfied, it can be reduced to a single eight order linear non-uniform resolving partial differential equation for the potential function $\Phi(x, y)$, with the help of which all accounting values are expressed.

Natural boundary conditions permit to determine internal force factors relations. Both forces and moments depend on longitudinal and tangential skin plane displacements functions $u_0(x, y)$, $v_0(x, y)$, and on deflection function $w(x, y)$. So it shows that the problem is integral; in other words, it can't be divided into plane part and bending part.

If the structurally-anisotropic composite panel structure derived as a result of energy averaging becomes orthotropic, linear differential operator of the resolving equation and linear differential operators of the relations between symmetrical stress-strain state components and potential function $\Phi(x, y)$ contain only even numbered derivatives with respect to x and y. Oblique symmetrical components are determined with odd numbered derivatives:

$$L^{(8)}\Phi = q\,, \tag{1}$$

$$L^{(8)} = \frac{K_{80}}{a^8}\frac{\partial^8}{\partial x^8} + \frac{K_{62}}{a^6 b^2}\frac{\partial^8}{\partial x^6 \partial y^2} + \frac{K_{44}}{a^4 b^4}\frac{\partial^8}{\partial x^4 \partial y^4} + \frac{K_{26}}{a^2 b^6}\frac{\partial^8}{\partial x^2 \partial y^6} + \frac{K_{08}}{b^8}\frac{\partial^8}{\partial y^8}$$

Constants $K_{8-i,i}$, i=0, 2, 4, 6, 8 in (1) depend on geometrical and stiffness structure characteristics; $x = x/a$, $y = y/b$ – non-dimensional coordinates with reference to the panel half-length a and to its width b.

General boundary value problem in the rectangular field is specified by elastic fixity conditions if:

$$\begin{aligned} x = const \quad & \left(\gamma_1 u_0 + \delta_1 N_x\right) = \left(\gamma_2 v_0 + \delta_2 N_{xy}\right) = \left(\gamma_3 w + \delta_3 \bar{Q}_x\right) = \left(\gamma_4 w'_x + \delta_4 M_x\right) = 0\,, \\ y = const \quad & \left(\alpha_1 u_0 + \beta_1 N_{yx}\right) = \left(\alpha_2 v_0 + \beta_2 N_y\right) = \left(\alpha_3 w + \beta_3 \bar{Q}_y\right) = \left(\alpha_4 w'_y + \beta_4 M_y\right) = 0\,, \end{aligned} \tag{2}$$

γ_i, δ_i, α_i, β_i , i = 1, 2, 3, 4 are from 0 to 1, which can be used to examine all possible boundary restraint combinations for connected plane and bending problems: free edge, simply supported edge, sliding fixation, constraining. In other words edge conditions may be kinematic, static and mixed.

Boundary value problems solution for (1) is constructed in a closed form by unitary Fourier series for particular boundary conditions along two opposite sides. These restrictions are called conformable one and they satisfy hinging condition about bending but sliding fixation condition in tangential direction for plane problem when the panel is loaded by shear force flows along its longitudinal edges. Face conditions are arbitrary and may be of elastic fixity one for symmetric or non-symmetric boundary value problems.

3. Statics problem results and discussion

As an example one calculated the displacements and stresses on the laminate macro-level in flat angle panel from carbon-plastic eccentrically stiffened in longitudinal direction. Edge conditions are symmetric by x coordinate. Normal loading $q(x, y) = const$.

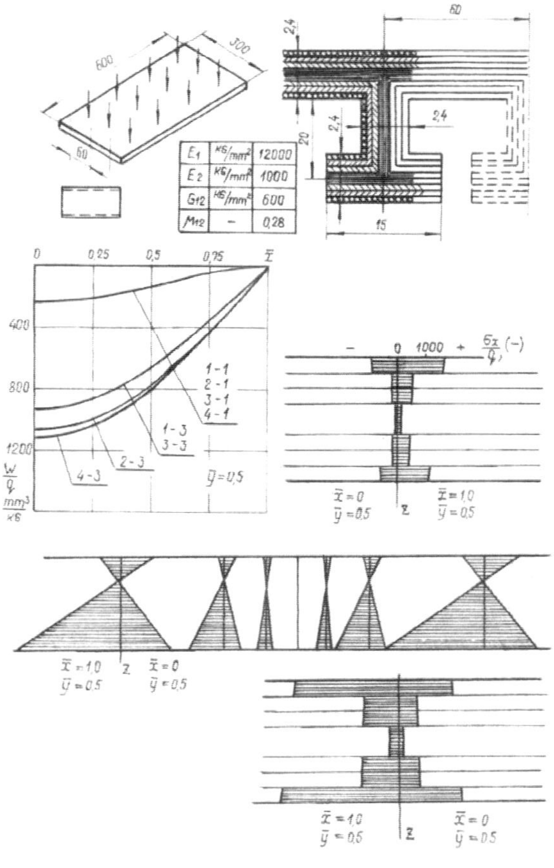

Figure 1. The panel, stiffened in longitudinal direction.
Deflections and normal stresses
Boundary conditions at lateral edges: plane problem – bending
1 – constraining, 2 – sliding fixation, 3 – simply supported edge, 4 – free edge

The deflections substantially consist on plane boundary conditions while lateral edges are simply supported about bending. When these edges are constrained for bending the deflection curves coincide practically while the restriction conditions about plane problem are varied.

As the next example we have calculated the residual thermal stresses brought about in cooling after hardening in flat angle panel made of carbon-epoxy having non-symmetric structure along its thickness (Figure 2). The panel contour is simply supported with respect to bending, and longitudinal sides restrictions correspond to the sliding fixation in tangential direction but lateral edges are free from forces and moments relative to plane problem.

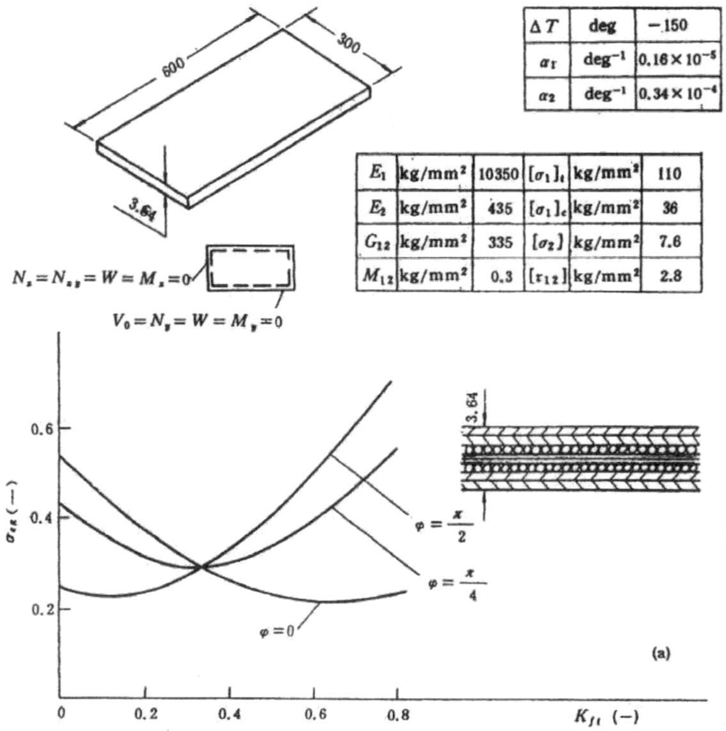

ΔT	deg	−150
α_1	deg^{-1}	0.16×10^{-5}
α_2	deg^{-1}	0.34×10^{-4}

E_1	kg/mm^2	10350	$[\sigma_1]_t$	kg/mm^2	110
E_2	kg/mm^2	435	$[\sigma_1]_c$	kg/mm^2	36
G_{12}	kg/mm^2	335	$[\sigma_2]$	kg/mm^2	7.6
M_{12}	kg/mm^2	0.3	$[\tau_{12}]$	kg/mm^2	2.8

Figure 2. The panel with non-symmetric package structure.
The influence of the reinforcing fibers preliminary tension on the residual temperature stresses

The estimation of the influence of the reinforcing fibers preliminary tension on the residual thermal stresses level has been performed. For the composite panel made of longitudinal, lateral and also oblique plies (Figure 2) the residual thermal stresses decrease while the reinforcing fiber pre-stressed tension increases. There is the optimum level K_{ft} = 0.35 (35% allowable *k*-ply fiber deformation) when the whole package stresses are equal and minimum.

4. Buckling problems

The papers [17–25] describe buckling problems of structurally-anisotropic composite panels but the presented mathematical model is seemed to be new one.

The differential equations of eighth order in partial derivatives are the resolving ones for the pre-critical stress-strained state and for the buckling problems [26].

The critical force definition of the general bending form of the thin-walled system buckling as well as the critical force definition of the many-waved torsion buckling are being of the most interest in accordance with traditional design practice. Both cases bending is integral with plane stress state so the buckling problem results in the value problem solving for the eight order partial derivatives equation in the rectangular field.

The solution is designed in closed form by double Fourier series for particular conformable boundary conditions along the panel contour. One can estimate the influence of production technology factors on the bearing strength of structurally-anisotropic composite panels if the before-critical stress state is taken into consideration or boundary conditions are non-conformable and the solution is formed by unitary trigonometric series or with the help of uniform solution method.

5. Buckling problem results and discussion

Figure 3 shows the influence of the composite panel sizes on the force level corresponding to general bending stability loss form and many-waved torsion buckling. The conditions of that or other buckling form existence are exposed.

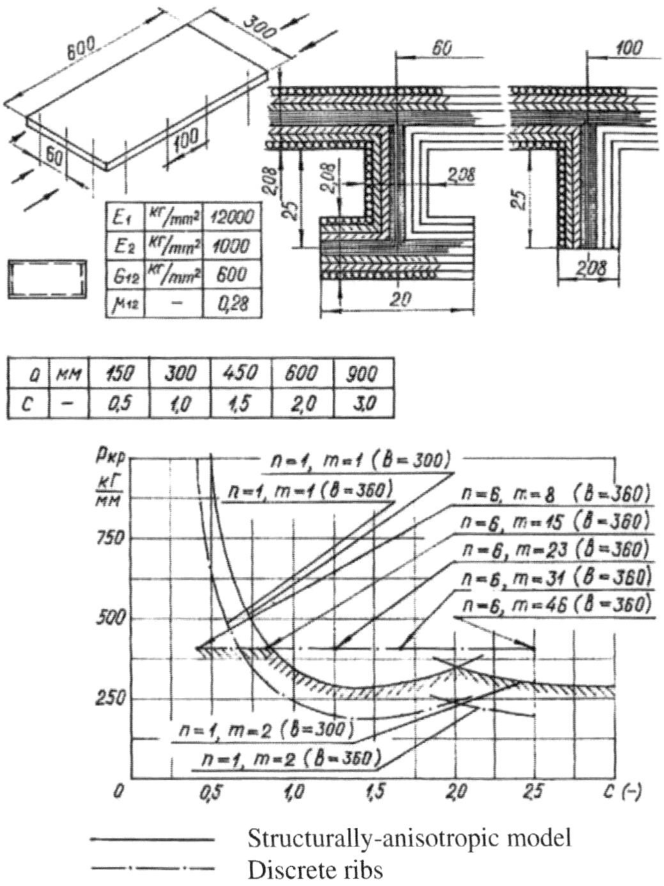

Figure 3. The panel with longitudinally-lateral stiffening set.
The influence of panel sizes on critical force level

For short panel if $c < 0{,}75$ there appears many-waved torsion buckling: $n = 6$, $m = 8$. If $c = 0{,}75$ the panel is equally buckled. The buckling for panels with $0{,}75 < c < 2{,}0$ has general bending form $n = 1$, $m = 1$, while $c > 2{,}0$ there is also general bending form, but $n = 1$, $m = 2$.

The testing series of uniform compressed stiffened composite panels carry to stability loss (Figure 4) have been made with the use of the special fixture.

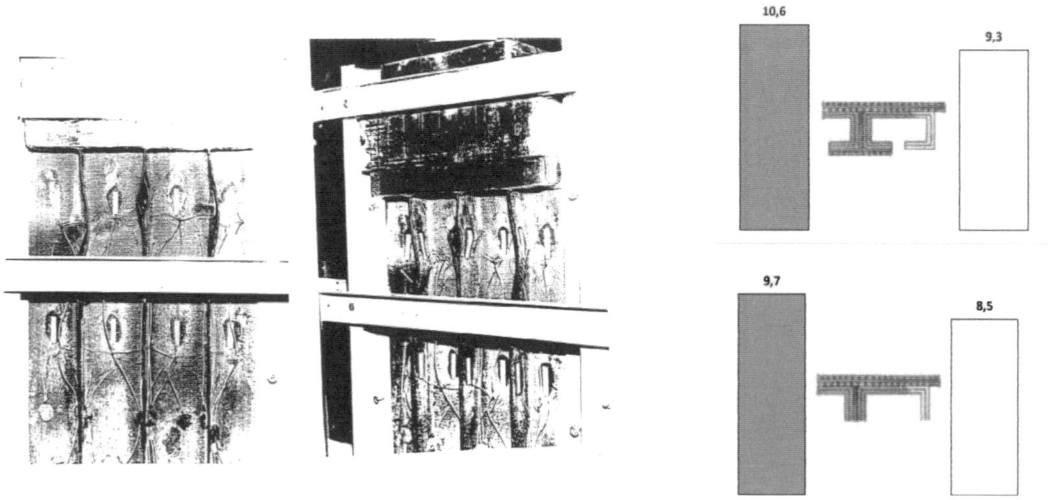

Figure 4. Experimental specimen

The refined theoretical results and the experimental data are in agreement qualitatively by buckling forms and quantitatively by critical stresses to the precision of 12% - 13%. It confirms the mathematical model authenticity.

6. Conclusion

Since we get the solution by analytical methods, the calculation time is of the minimum one. That is of interest from the point of view of practical design using parametric analysis. The results of the stress analysis calculations as well as the results of the buckling analysis calculations offer the opportunities for reducing and optimization of the aircraft elements weight characteristics.

References

[1] Adumitroaie A and Barbero E J 2015 *Mech. Adv. Mater. and Struct.* **22** (9) 705
[2] Auricchio F, Balduzzi G, Kroshgoftar M J et al 2014 *Compos. Struct.* **118** 6223
[3] Boutin C and Viverge K 2016 *Eur. J. Mech. A* **55** 149
[4] Cerracchio P, Gherlone M and Tessler A 2015 *Meccanica* **50** (10) 2487
[5] Celkovic M 2015 *Compos. Struct.* **125** 388
[6] Fernandes G R, Denipotti G J and Konda D H 2010 *Comput. Mech.* **45** (6) 523
[7] Goodsell J, Pagano N J et al 2013 *Trans. ASME. J. Appl. Mech.* **80** (4) 041020
[8] Kant T and Shiyekar S M 2013 *Compos. Struct.* **96** 698
[9] Le A L, Nguyen T T, Ho H V, Dang T H and Bui X T 2015 *Compos. Struct.* **127** 382
[10] Li Y B, Zhang P, Wu S Q and Fei Q G 2015 *J. Vibr. Eng.* **28** (4) 531
[11] Luan Y, Ohlrich M and Jacobsen F 2011 *J. Sound and Vibr.* **330** (17) 4274
[12] Mantan J L and Guedes S C 2014 *Int. J. Mech. Sci.* **78** 60
[13] Nath S K D and Afsar A M 2009 *Mech. Adv. Mater. and Struct.* **16** (4) 300
[14] Petrolo M and Lamberti A 2016 *Mech. Adv. Mater. and Struct.* **23** (1) 28
[15] Tran L V, Nguyen T T , Thai C H and Nguyen X H 2015 *Mech. Adv. Mater. and Struct.* **22** 248
[16] Boytsov B V and Gavva L M 2017 *J. Quality and life* **2** 84 (In Russian)
[17] Chen C S, Lin C Y and Chien R D 2011 *Int. J. Mech. Sci.* **53** (1) 51
[18] Huang L, Sheih A H, Ng C T and Griffin M C 2015 *Compos. Struct.* **122** 41

[19] Kazemi M 2016 *Arch. Appl. Mech.* **85** (11) 1667
[20] Kim C, Kim Y K and Choi W 2002 *Trans. Kor. Soc. Mech. Eng. A* (8) 1653
[21] Lindgaard E, Lund E and Rassmussen K 2010 *Int. J. Solids and Struct.* **47** 3186
[22] Mittelstedt C and Schroder K U 2010 *Compos. Struct.* **92** 2830
[23] Shafei E and Kabir M Z 2011 *Appl. Compos. Mater.* **18** (6) 539
[24] Vescovini R and Dozio L 2015 *Compos. Struct.* **127** 356
[25] Yeter E, Erklig A and Bulut M 2014 *Compos. Struct.* **118** 19
[26] Firsanov V V and Gavva L M 2017 *Structural Mechanics of Engineering Constructions and Buildings* **4** 66 (In Russian)

IOP Conf. Series: Materials Science and Engineering **312** (2018) 012010 doi:10.1088/1757-899X/312/1/012010

Nonparametric method for failures detection and localization in the actuating subsystem of aircraft control system

S S Karpenko, E Yu Zybin and V V Kosyanchuk

State Research Institute of Aviation Systems, Moscow, Russia

kss@gosniias.ru

Abstract. In this paper we design a nonparametric method for failures detection and localization in the aircraft control system that uses the measurements of the control signals and the aircraft states only. It doesn't require a priori information of the aircraft model parameters, training or statistical calculations, and is based on algebraic solvability conditions for the aircraft model identification problem. This makes it possible to significantly increase the efficiency of detection and localization problem solution by completely eliminating errors, associated with aircraft model uncertainties.

1. Introduction

The most dangerous failures in the aircraft control system are the failures of the elements of the actuating subsystem of aircraft control system. For example: the failure of the actuator control unit, the failure of the communication link between the control unit and the actuator, the failure of the actuator itself, damage of the control surfaces. In the event of such failures, the aircraft aerodynamic coefficients and the control surfaces moment characteristics are changed, that can lead to an aviation accident. If the failure detection time exceeds the critical control system reaction time, the aircraft can move to nonrecoverable state. Therefore, the primary task is to detect of aircraft abnormal dynamics and localize the failure before an emergency. Fast failures detection and localization in the aircraft allows minimizing their consequences and taking early measures to prevent an accident.

The failures detection and localization methods can be classified into two broad groups: parametric and nonparametric [1–5]. As a rule, parametric methods are used to detect and localize failures, directly or indirectly implying a priori information about the parameters of the aircraft model. Using the parametric model-based methods involves a number of difficulties due to the nonlinearity of such models, inaccuracy in determining their parameters, multiply solutions, etc. The model parameter errors may often increase the threshold values of the failure detection criteria, thus increasing the time of failure detection and decreasing the accuracy of determining the time the failure occurred. In some cases, the accurate determination of the time that the failure occurred proves to be very difficult and time consuming [6–8].

In contrast to the parametric methods, the nonparametric methods of failures detection and localization based only on measurement of input and output signals and need no a priori information about model parameters [9–10]. These methods are based on the determination of specific qualitative or quantitative model characteristics. The all widely known nonparametric methods either require preliminary training/tuning for a particular plant or use the statistical algorithms, that themselves are subject to inevitable errors. Determining accurate and reliable solutions using such algorithms requires a large amount of data. They are characterized by high computational costs and low performance indicators. These factors determine the high probability of false warning or missed failures in practice [11–14].

The proposed method for aircraft control system failures detection and localization is based on the measurements of the control signals and the aircraft states only. It doesn't require a priori information of the aircraft model parameters, training or statistical calculations, and is based on algebraic solvability conditions for the aircraft model identification problem. This makes it possible to significantly increase the reliability and speed of problem solution by completely eliminating parametric errors.

2. Statement and solution of failures detection and localization problem

Let the model of a non-failure aircraft dynamics be represented in the state space as [15]

$$x_{i+1} = Ax_i + Bu_i,\ x^f_{j+1} = Ax^f_j + BFu_j, \tag{1}$$

where A, B are the matrices of eigen-dynamics and control efficiency; x, u are the state and control vectors of length n_x, n_u; $i = \overline{0, l-1}$, $j = l, l+1, \ldots$ are the discrete times before and after the occurrence of failures; l is the instant a failures occur; $F = \operatorname{diag}\left[f(1) \cdots f(k) \cdots f(n_u)\right]$ is the matrix of failures (loss of efficiency), $f(*) = 1$ for a non-failure control channel, $f(*) = 0$ for a failure control channel. It is necessary, based on the measurements of control signals u and states x only, to detect the failures and localize the failure channels of the aircraft control system.

Suppose that observation of aircraft control signals and states is conducted for some time. Then the aircraft models (1) will have the matrix forms:

$$X_{i+1} = AX_i + BU_i,\ X^f_{j+1} = AX^f_j + B_f U_j, \tag{2}$$

where $B_f = BF$; $X_i = [x_i \ \ldots \ x_{i+h}]$, $X^f_j = \left[x^f_j \ \ldots \ x^f_{j+h^f}\right]$, $U_i = [u_i \ \ldots \ u_{i+h}]$, $U_j = \left[u_j \ \ldots \ u_{j+h^f}\right]$; h, h^f are the number of observations in non-failure and failure cases. Let's represent the expressions (2) in the form of the following matrix equations:

$$[-A \ I]\begin{bmatrix} X_i \\ X_{i+1} \end{bmatrix} = BU_i,\ [-A \ I]\begin{bmatrix} X^f_j \\ X^f_{j+1} \end{bmatrix} = B_f U_j. \tag{3}$$

It's known [16], that any linear matrix equation of the form

$$YC = D$$

with known matrices C, D is solvable for Y if and only if the solvability condition

$$D\overline{C}^R = 0, \tag{4}$$

is satisfied, where $\overline{C}^R$ is the right zero divisor of maximal rank (matrix for which condition $C\overline{C}^R = 0$ is satisfied).

As the parameters of free aircraft dynamics don't change at emergence of failures in aircraft control system, for the solvability of (3) in accordance with (4) the following conditions must be satisfied:

$$BU_i\overline{\begin{bmatrix} X_i \\ X_{i+1} \end{bmatrix}}^R = 0,\ B_f U_j\overline{\begin{bmatrix} X^f_j \\ X^f_{j+1} \end{bmatrix}}^R = 0. \tag{5}$$

If there is no functional redundancy in the actuating subsystem of aircraft control system $(\bar{B}^R = 0, \bar{B}_f^R = 0)$ the conditions (5) have an equivalent form, which does not depend on the parameters of the aircraft control efficiency:

$$U_i \overline{\begin{bmatrix} X_i \\ X_{i+1} \end{bmatrix}}^R = 0,\ U_j \overline{\begin{bmatrix} X_j^f \\ X_{j+1}^f \end{bmatrix}}^R = 0. \tag{6}$$

But at the time of failures occur the solvability condition does not hold and looks like

$$\begin{bmatrix} U_i & U_j \end{bmatrix} \overline{\begin{bmatrix} X_i & X_j^f \\ X_{i+1} & X_{j+1}^f \end{bmatrix}}^R = \left(B_f - B\right)\begin{bmatrix} 0 & U_j \end{bmatrix} \overline{\begin{bmatrix} X_i & X_j^f \\ X_{i+1} & X_{j+1}^f \end{bmatrix}}^R \neq 0. \tag{7}$$

Having viewed (7) line by line, it can be seen that for the failure control channels ($f(*) \neq 1$) the solvability conditions are not satisfied, and for the non-failure ($f(*)=1$) – are satisfied. This allows us to use the norms of conditions for the k of n_y control channels $U_* = \left[u_*^{(1)} \ldots u_*^{(k)} \ldots u_*^{(n_y)}\right]^{\mathrm{T}}$

$$\sigma^{(k)} = \left\| \begin{bmatrix} u_i^{(k)} & u_j^{(k)} \end{bmatrix} \overline{\begin{bmatrix} X_i & X_j^f \\ X_{i+1} & X_{j+1}^f \end{bmatrix}}^R \right\|_2 \tag{8}$$

as nonparametric criterion for aircraft control system failures detection and localization.

3. Aircraft control system failures detection and localization example

Let's assume that for one of the aircraft flight modes, the matrices of the eigen-dynamics and control efficiency of aircraft model (1) with in-phase deviation of elevator rudders and differential aileron deflection have the form:

$$A = \begin{pmatrix} 0,9998 & -0,1021 & -0,0981 & 0,0000 & 0,0000 & 0,0000 & 0,0000 & 0,0000 \\ -0,0001 & 0,9907 & 0,0000 & 0,0100 & 0,0000 & 0,0000 & 0,0000 & 0,0000 \\ 0,0001 & 0,0093 & 0,9999 & 0,0000 & 0,0000 & 0,0000 & 0,0000 & 0,0000 \\ 0,0000 & -0,0079 & 0,0000 & 0,9973 & 0,0000 & 0,0000 & 0,0000 & 0,0000 \\ 0,0000 & 0,0000 & 0,0000 & 0,0000 & 1,0003 & 0,0011 & 0,0099 & 0,0014 \\ 0,0000 & 0,0000 & 0,0000 & 0,0000 & -0,0344 & 0,9650 & -0,0090 & 0,0000 \\ 0,0000 & 0,0000 & 0,0000 & 0,0000 & -0,0199 & 0,0020 & 0,9896 & 0,0000 \\ 0,0000 & 0,0000 & 0,0000 & 0,0000 & 0,0000 & 0,0100 & -0,0011 & 1,0000 \end{pmatrix},$$

$$B = \begin{pmatrix} 0,00000 & 0,00000 & 0,00000 & 0,00000 \\ 0,00036 & -0,00008 & 0,00000 & 0,00000 \\ -0,00036 & 0,00008 & 0,00000 & 0,00000 \\ -0,00676 & -0,01343 & 0,00000 & 0,00000 \\ 0,00000 & 0,00000 & -0,00005 & 0,00000 \\ 0,00000 & 0,00000 & -0,00197 & -0,01084 \\ 0,00000 & 0,00000 & -0,03146 & 0,00116 \\ 0,00000 & 0,00000 & 0,00000 & 0,00000 \end{pmatrix}.$$

In this case, the vectors of states and control are written as $x=\left[V\ \alpha\ \vartheta\ \omega_z\ \beta\ \omega_x\ \omega_y\ \gamma\right]^{\mathrm{T}}$, $u=\left[\delta_{\mathrm{el}},\delta_{\mathrm{stab}},\delta_{\mathrm{rud}},\delta_{\mathrm{ail}}\right]^{\mathrm{T}}$, where $\omega_x,\omega_y,\omega_z$ are the angular rates of roll, yaw, and pitch (deg/s); $\alpha,\vartheta,\gamma,\beta$ are the angles of attack, pitch, roll, and slip (deg); V is the flight speed (m/s); $\delta_{\mathrm{el}},\delta_{\mathrm{stab}},\delta_{\mathrm{rud}},\delta_{\mathrm{ail}}$ are the angles of elevators, stabilizer, rudder and ailerons deflections (deg).

Let us simulate the flight of the aircraft during 40 s with the following control signals: $\delta_{\mathrm{el}}=5\sin(0{,}05\cdot t)$, $\delta_{\mathrm{stab}}=2\sin(0{,}025\cdot t)$, $\delta_{\mathrm{rud}}=-5\sin(0{,}1\cdot t)$, $\delta_{\mathrm{ail}}=5\cos(0{,}05\cdot t)$. Failures in the form of loss of efficiency of each control channel by 10% $(f(*)=0{,}9)$ will be entered on the 20th second of the flight $(i=0:19,\ j=20:40)$.

The determination of the failure control channels is carried out by the failures detection and localization criterion (8) with $h=h^f=13$, the graphs of which for each failure case are shown in figure 1.

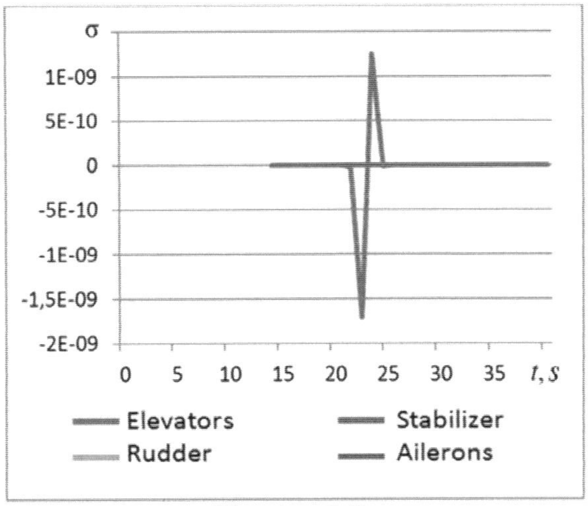

Elevators failure

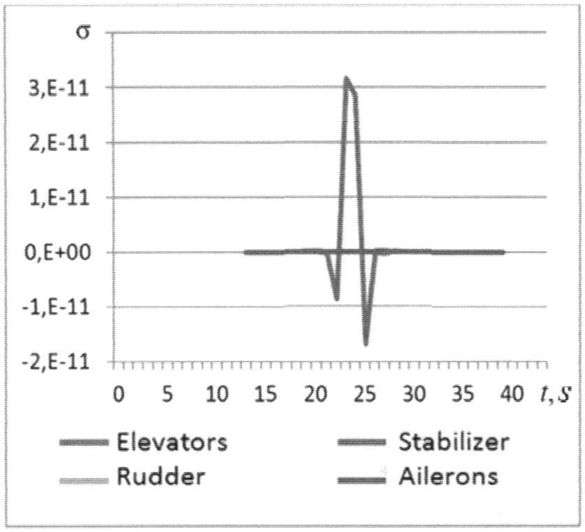

Stabilizer failure

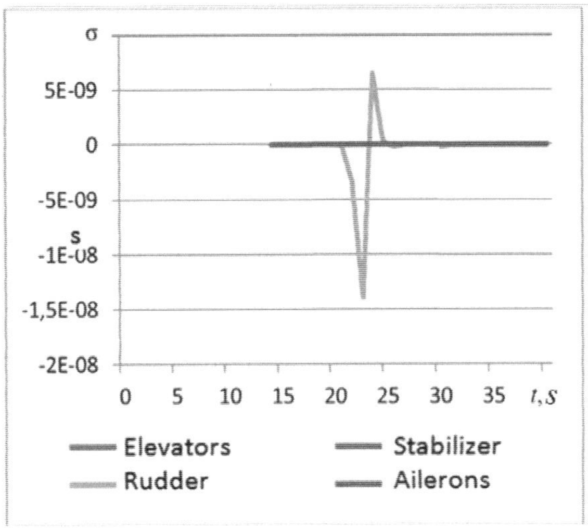

Rudder failure

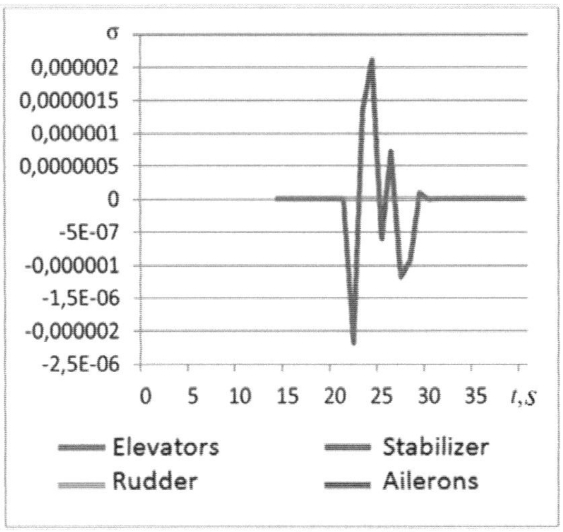

Ailerons failure

Figure 1. Graphs of failures detection and localization criterion.

It is clear that in each case of failure, its detection and localization are carried out fast and reliably.

4. Conclusion

As a result the new method of failures detection and localization in aircraft control systems, using only data of control signals and state measurements, is developed. The main advantage of it is its independence from the aircraft model parameters, which guarantees its efficiency in the absence of a priori information, does not require solving the problems of identifying model parameters and predicting its dynamics.

Acknowledgments

This work is supported by Russian Foundation for Basic Research (17-08-01445-a, 17-01-00251-a)

References

[1] Patton R J 1997 *Proc. of the IFAC Symposium on Fault Detection Supervision and Safety for Technical Processes* 1033–1055
[2] Zhang Y and Jiang J 2008 *Ann. Rev. Control* **32** (2) 229–252
[3] Isermann R 2006 *Fault-diagnosis Systems: an Introduction from Fault Detection to Fault Tolerance* (Berlin-Heidelberg: Springer) p 475
[4] Venkatasubramanian V, Rengaswamy R, Kavuri S and Yin K 2003 *Computers & Chemical Engineering* **27** (3) 293–311
[5] Venkatasubramanian V, Rengaswamy R, Kavuri S and Yin K 2003 *Computers & Chemical Engineering* **27** (3) 313–326
[6] Zybin E Y and Kos'yanchuk V V 2016 *Journal of Computer and Systems Sciences International* **55** (4) 546–557
[7] Korsun O N 2008 *Mekhatronika, Avtomatizatsiya, Upravlenie* S**6** 2–7
[8] Korsun O N, Om M H, Latt K Z and Stulovskii A V 2017 *Procedia Computer Science* **103** 67–74
[9] Venkatasubramanian V, Rengaswamy R, Kavuri S and Yin K 2003 A *Computers & Chemical Engineering* **27** (3) 327–346
[10] Zhirabok A, Pavlov S and Shumsky A 2016 *Proceedings of the IECON 2016* 429–434
[11] Hwang I, Kim S, Kim Y and Seah C E 2010 *IEEE Transactions on Control Systems Technology* **18** (3) 636-653
[12] Meskin N and Khorasani K 2011 *Fault detection and isolation: multi-vehicle unmanned systems* (New York: Springer) p 166
[13] Zolghadri A 2013 *EDP Sciences* 231–248
[14] Anupam D, Maiti J and Banerjee R P 2012 *Intern. J. Quality & Reliability Management* **29** (7) 720–752
[15] Zybin E, Kosyanchuk V and Karpenko S 2017 *MATEC Web of Conferences* **99** 03011
[16] Zybin E Yu, Kos'yanchuk V V, Kul'chak A M and Sel'vesyuk N I 2016 *GosNIIAS Transaction. Series: Avionics* **2** (26) 3–11

IOP Conf. Series: Materials Science and Engineering **312** (2018) 012011 doi:10.1088/1757-899X/312/1/012011

Morphology and kinetics of the formation of polyamide benzimidazole in N-methylpyrrolidone

O A Khanchich [1,2*], V B Larionov[1] and A L Biryukov[2]

[1] Moscow Technological Institute, Moscow, 119334, Russia
[2] RGAU-MSXA K.A. name Timiryazev, Moscow, 127550, Russia

*E-mail: o_hanchich@mti.edu.ru

Abstract. Polymerization-optical methods were used to study the supermolecular structure in the gelation of systems based on polyamidobenzimidazole (PABI) in N-methylpyrrolidone (N-MP) and the kinetics of phase transformations during polymer coagulation from a the solution. It was found that the use of N-MP leads to the appearance in the PABI-N-MP-H2O system of spherulites having a structure characteristic of LC structures. It is shown that the alternation of metastable structures upon gelation of the PABY – N-MP-H2O system is consistent with the appearance of the LC phase of the cholesteric type.

1. Introduction

At present, the third generation of reinforcing fibers has become extremely important. Traditionally used materials (metals, metal alloys) are increasingly being replaced with new ones – composites, the production of which is impossible without reinforcing fibers. These include heterocyclic p-amide fibers of CBM and Armos thet have the highest mechanical properties among the entire family of n-aramid fibers [1]. Such fibers are obtained from concentrated sulfuric acid solutions of polyamide benzimidazole (PABI). The study of the morphology of PABI-based systems in other types of solvents, the kinetics of phase transformations with slow gelation of the polymer-solvent-precipitant system, which are in a jelly-like state, is of independent scientific interest. On such systems, it is possible to experimentally test the methods for controlling the stiffness of macromolecules by changing the type of solvent, as well as by increasing the polymer concentration both directly in solution, and when polymer is precipitated from the solution. The latter implies the possibility of the system to possess the properties of a lyotropic liquid crystal. However, the achievement of a critical concentration of the transition to the anisotropic phase in this case should occur directly during the coagulation process.

The aim of this work is to study the morphology of PABI based systems in other types of solvents, the kinetics of phase transformations during polymer coagulation from solution, and also during slow gelation of the polymer-solvent-precipitant system, which are in a gelatinous state. Taking into account the rigid-chain nature of PABI, it can be expected that during its gelation or precipitation from solutions in other solvents, different structures will be formed, characteristic of both rigid-chain and flexible-chain polymers. This follows from Flory's theory [2], which associates the critical concentration of the transition to the LC state (C*) with the degree of anisometry of macromolecules. PABI in N-methylpyrrolidone (N-MP) with the length (nm) of the Kuhn statistical segment A = 530 nm was selected as objects of investigation. The ability of polyheteroarylenes to form an LC phase in the same solvents may be a relative stiffness criterion. For such polymers, this can be sulfuric acid. From this point of view, poly-*n*-benzamide (C* = 7 - 8%) and poly-*n*-phenylene terephthalamide (C* = 8 - 9%) should be considered the most rigid-chain of this range of polymers [3]. Further, in the series of decreasing rigidity there are PABI (C* = 11 - 13%) [4].

2. Experimental

Let us consider the kinetic features and structure of the brittle PABI, depending on the temperature and concentration conditions of gelling in the N-MP, containing 1.3% of water. For the preparation of solutions, unfractionated PABI samples were used, obtained by isolating the polymer in the form of films from a solution with an intrinsic viscosity [η] = 5.8 dl/g, measured at 20°C in dimethylacetamide. The solutions were kept in sealed cuvettes with a gap of 0.5 mm (Figure 1).

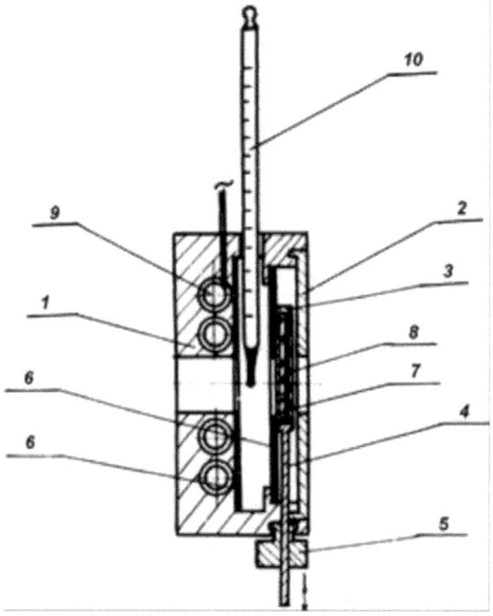

Figure 1. Scheme of the device for testing solutions and films of polymers: 1 - thermostat, 2 - thermostat cover, 3 - sample chamber, 4 - screw displacement, 5 - nut rotation, 6 - asbestos gaskets, 7 - sample, 8 - coverslips, 9 - heating element, 10 - thermometer.

The transformation of a polymer solution into a jelly during coagulation in a precipitation bath is a kinetic process, associated with structural changes that often occur within a fraction of a second. This circumstance, as well as a number of other factors, cause certain experimental difficulties during the study of the kinetics of structure formation by, for example, such methods as electron microscopy and x-ray diffraction analysis. It is known that the most important property of optically anisotropic supermolecular structures is their ability to scatter and depolarize visible light [5]. Therefore, the main methods of investigation were polarization-optical methods. According to H_v (crossed polaroids) and V_v -scattering patterns of polarized light (parallel polaroids), it is also possible to determine the type of formed mesophases [6]. Diffractograms describing the changes in the structure of the object during coagulation were obtained at the installation (Figure 2), which was designed and manufactured on the basis of the general optical scheme for small-angle polarized-light-scattering (SALS).

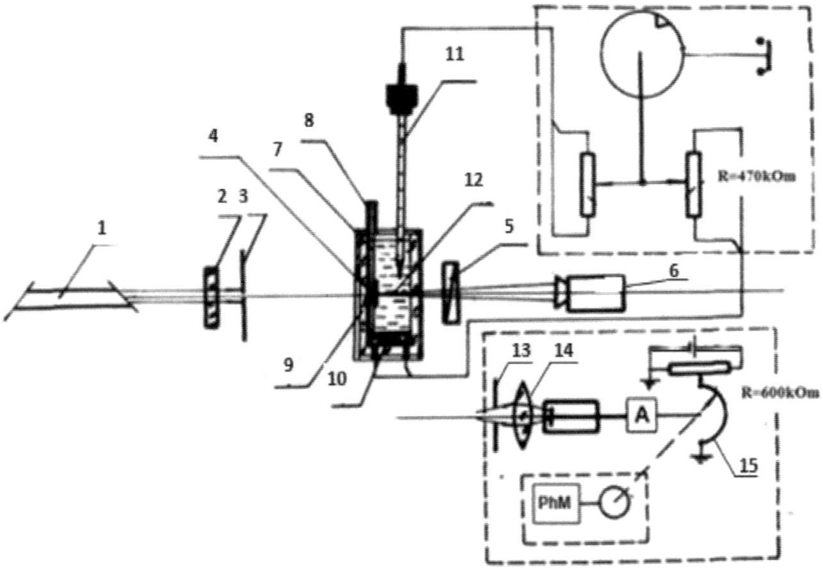

Figure 2. Scheme of experimental setup of the MRPS for study of the kinetics of rapidly-occuring structure formation processes in the precipitation of polymer from solution: 1 – laser, 2 – interference filter, 3,13 – diaphragms, 4 – specimen, 5 – analyzer, 6 – motion picture camera, 7 – object glass, 8 – guide, 9 – cuvette, 10 – heating elements, 11 – temperature sensor, 12 – precipitation bath, 14 – collecting lens, 15 – slide wire, PhM - photomultiplier.

A relatively small deterioration in the quality of the solvent leads to an increase in viscosity and gelation of the solution over time (Figure 3).

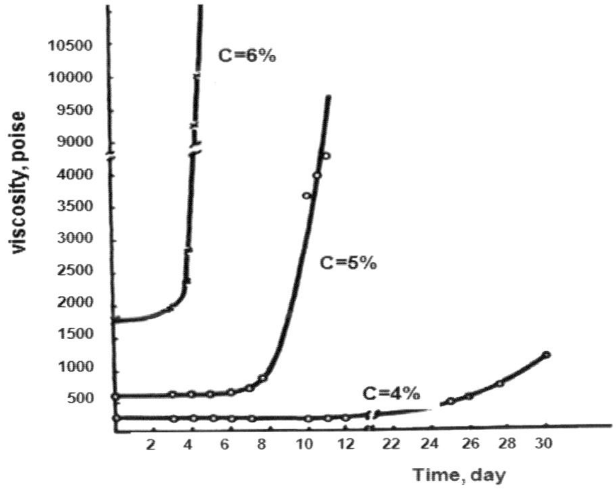

Figure 3. Change in viscosity of PABI solutions of various concentrations containing 1.8% water in N-MP, from the time of gelation.

At the same time, the transparency of the solution also changes, with solution C = 4% becoming turbid in 21 days, C = 5% in 11 days and C = 6% in 4 days. The change in the degree of turbidity of the solution is not accompanied by the appearance of any structural elements registered in sealed cuvettes by the SALS method. On the diffractometer operating in the photometer mode (Figure 2) an increase in the overall anisotropy of the system is observed. This corresponds to aggregation at the molecular level (below the Rayleigh-Hans resolution region)

$$4\pi R(n'-1)/\lambda << 1,$$

where R – is the radius of the particle, n' - is the refractive index of the particle relative to the medium. In the polymer system, n' is close to unity and the Rayleigh-Hans approximation covers a wide range of particles (inhomogeneities) in size (experimentally, from 0.1 to 20 μm).

Visible changes in scattering patterns are manifested only when very high viscosity solutions (more than 10^5 poises) are reached. Solutions with a polymer concentration of less than 4%, despite a significant increase in viscosity and the appearance of small spherulites recorded by light scattering, do not form strong jelly for several months. And the transition of the system to the equilibrium state occurs through several stages of structuring and is accompanied by a variety of supramolecular structures (Figure 4).

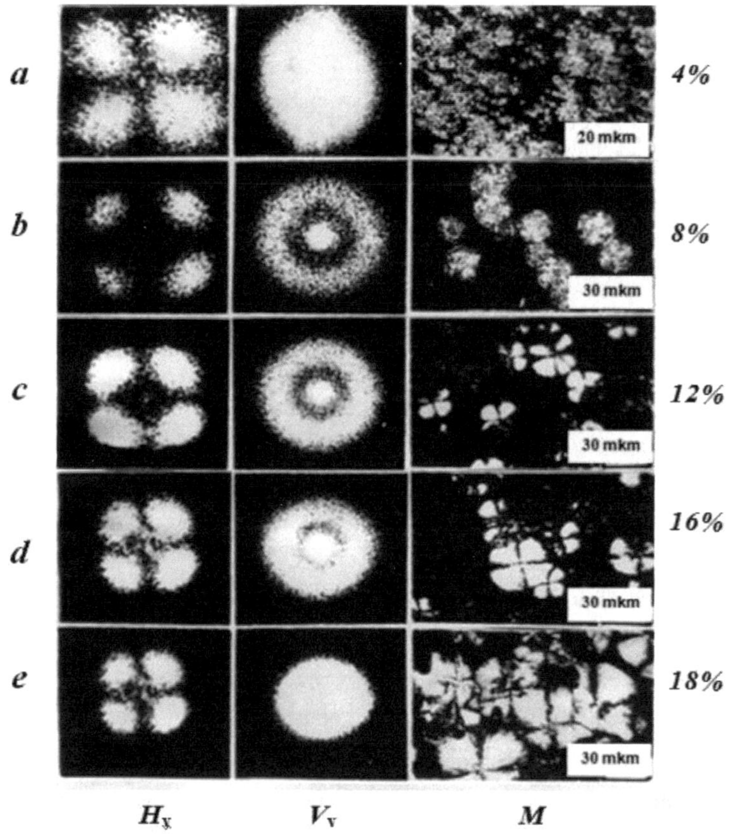

Figure 4. H_v and V_v of the light scattering pattern and microphotography (M) of PABI solutions of different concentrations in two months after preparation.

The appearance of small spherulites in the system depends on the concentration of the polymer. In 4% solutions, the spherulite structure is formed 2 months after the solution is prepared (Figure 4, a). The distribution of the polarizability ellipsoids (crystallites), as can be seen from scattering patterns, is "correct" and is for radial spherulites with dimensions of up to 3.5 mkm. At large concentrations of the solution, large spherical particles (up to 30 mkm) arise, the "assembly" of which occurs from the ordered material of the original spherulite structures (Figure 4, b). The optical properties of such particles differ from spherulites. In H_v, the scattering pattern gives rise to discrete reflexes, which, upon transition to V_v, merge into a ring. From the analysis of these scattering patterns it can be concluded that the crystallites in spherical particles are oriented arbitrarily and are connected to each other in a coil by an amorphous matrix.

With increasing time and concentration, self-ordering takes place in these particles with the formation of large spherulites that exhibit a bright interference color of the quenching patterns (*M*) in the form of "maltese crosses" (Figure 4, c). Gradually, these structures increase in size (Figure 4, d) and at 18% fill the entire field (Figure 4, e). After 2.5 months in a 3% solution, during a certain period of time (10-13 days) three types of structures coexist simultaneously: small spherulites, spherical structures and large spherulites, "colored" in crossed nicols of a polarizing microscope. These spherulites are formed from separate ordered coils. Photomicrographs in the crossed Nicol prisms of a polarizing microscope show a change in the color of the formed spherically symmetric "prefabricated" structures with a radial-ray structure and sizes up to several microns (Figure 5).

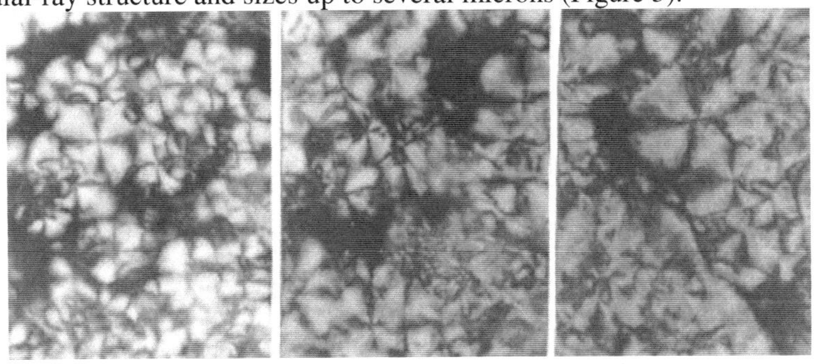

(a) (b) (c)

Figure 5. Microphotographs (nicols crossed) of 12% jelly PABI in N-MP at gelling times (months): a-1,5; b-2,5; c-3,5. Magnification 300^x

The change in the interference color over time and the appearance of brighter tones indicate an internal restructuring of the spherulites. The increase in gelation time or polymer concentration in the solution is accompanied by an increase in the size and number of scattering structures in the jelly. The intensities H_v and V_v of scattering patterns become comparable in magnitude (Figure 4, d). After 5 months in 16% solutions, the number and size of the spherulites formed become so large that their collapse occurs with the appearance in the system of disorderly orientation of the scattering elements. From the point of view of the SALS theory, this indicates a predominance of orientation fluctuations in the scattering medium, which is qualitatively similar to light scattering by polymer films with a correlation structure [5]. At the same time, if the concentration of solutions is less than 2%, they only change transparency. This allows us to consider such systems as nonequilibrium, in which the establishment of thermodynamic equilibrium requires more time than in concentrated solutions.

3. Results and discussion

Polarization-optical studies of the jellied PABI in N-MP show their analogy with low-molecular-weight cholesteric liquid crystals. In particular, the scattering of light in a thin layer of cholesterol oleate is similar to the diffractograms in Figure 4, c. The "maltese crosses" observed in the microscope correspond to the confocal texture of the liquid crystal. When the low-molecular-weight cholesteric LC layer is shifted, a transition to a homogeneous planar texture occurs. The scattering pattern in this case is identical to the diffraction pattern observed when the PABI is deposited under "soft" conditions (precipitant is a 40% aqueous solution of N-MP). Diffractograms of this type correspond to a rod-like structure with orientation of the main direction of polarizability along the axis of the rod.

The alternation of metastable structures during gelation of the PABI–N-MP–H_2O system is also consistent with the previously described occurrence of the LC phase of the cholesteric type from the spherulite structure in the process of isothermal structuring in solutions of poly-γ-benzyl-*L*-glutamate [6]. Our crystal-optical experiments allow us to assume that the structure of spherulites of rigid-chain polymers of the PABI type differs significantly from spherulites in flexible-chain polymers. This is also confirmed by the absence of a crystalline phase in the spherulite-containing studs of PABI when they are examined by x-ray diffraction. The difference between the supramolecular structure of flexible-chain and rigid-chain polymers is that for the former the folded form of the macromolecule and the formation of crystals with folded chains during crystallization is thermodynamically preferred. Their thickness is many times smaller than the length of the straightened macromolecule. It is known that in flexible-chain polymers, crystalline regions alternate with amorphous sections in the radial or tangential directions [7].

The spherulites observed in the form of brightly colored "maltese crosses" in the PABI studios can be represented as consisting of ordered macromolecules. They are distributed along the radius, and are interconnected not by an amorphous interlayer but by ordered units having a two-dimensional order at the molecular level corresponding to the liquid crystal structure. The coloring of quenching patterns is determined by the specifics of the internal structure of the PABI spherulites. The improvement of such spherulites during gelling, observed as the decomposition of the light beam into interference components, occurs by the continuous addition of parallel links of macromolecules during the gelation process. Confirmation that the PABI under certain conditions is capable of forming a cholesteric mesophase with a large spiral twist pitch $P >> \lambda$ is indicated by experiments on the gelation of this polymer in other dipolar aprotic solvents.

References

[1] Perepelkin K E 2009 *Reinforcing fibers and fibrous polymeric composites* (Moscow, Scientific Foundations and Technologies Publishing House)
[2] Flory P J 1956 *Proc. Roy. Soc* **234** 60-72
[3] Kulichikhin VG, Vasil'eva NV, Platonov VA et al 1976 *J. High-molecular Compounds A* **18** 590-596
[4] Khanchich OA and Petrova LV 1983 *Abstracts of the reports of the All-Union Scientific and Technical Conference* "Theory and Practice of Chemical Fibers Formation" pp 191-193
[5] Stein R S, Erhardt P, van Aartsen J J, Clough S and Rhodes M B 1966 *J. Polym. Sci.* **13** 1-36
[6] Frenkel S Ya 1974 J. Appl. Chem. **38** 1-17
[7] Takase A, Sakagami S, Nakamiso M Japan 1973 *J. Appl. Phys.* **12** 1255-1261

IOP Conf. Series: Materials Science and Engineering **312** (2018) 012012 doi:10.1088/1757-899X/312/1/012012

Technical diagnostics functioning machines and mechanisms

M I Kiselev[1], V I Pronyakin[1] and A K Tulekbaeva[2]

[1]Bauman Moscow State Technical University, Moscow, 105005, 5, Russia
[2]M. Auezov South Kazakhstan State University, Shymkent, 160012, Republic of Kazakhstan

E-mail: vip-u@yandex.ru

Abstract. Article discusses the machines and mechanisms technical state monitoring problem. Approaches for estimating mechanical systems current technical state, defects detection and evaluation of mechanical elements degradation levels are considered. The paper analyzes the traditional methods offered in international and national standards, especially vibrodiagnostics. An advanced phase method is presented which is based on registration the kinematic parameters of the mechanism running cycle. The result of coupling the phase method and mathematical modeling is shown, and simulation comparison with the experimental data is presented.

1. Introduction

Machines and mechanisms technical state monitoring is an important part of maintenance process. The principal tasks are the current state monitoring, the degradation degree evaluation and emergency protection.

The current state of art is given in international standards, also applied in Russia [1-5]. These standards provide general approaches, methods, procedures and methodology used to determine the strategy of monitoring the current state of functioning machines and mechanisms, as well as the interpretation of measurement results. The data processing includes the parametric diagnostics , for example, vibration parameters, temperature, power, operating speed, contaminant content.

Approaches to monitoring consist in the analysis of a design, identification of diagnostic signs at faults and breakages and tracing the links between cause and effect which led to defect.. Also the operating experience of the similar equipment is used. It is acknowledged that when using traditional approaches it is very hard to obtain quantitative assessment of degradation degree of cyclic mechanism and to provide the emergency protection.

The monitoring and diagnostics strategy main stages are:

- identification of the most critical components of machines and mechanisms,
- analysis of failures during operation,
- forming the list of measured parameters required for monitoring,
- faults and defects diagnostic signs identification,
- monitoring strategy development.

For development of monitoring strategy the operating experience for long periods of time is required in order to accumulate the score for defects and results of their analysis. And only after that the development of monitoring and diagnostics systems becomes possible. However not all defects are manifested at the initial stage of operation. Such long preparatory process is unacceptable for the expensive equipment at high costs of its maintenance.

The analysis of methodology of diagnostics stated in the international and national standards shows that the level of obtained information is not insufficient for reliable assessment of functioning of machines and mechanisms. For traditional approaches the insufficient level of information is characterized by lack of quantitative estimates and wide use of qualitative categories for defect probability detection: "low, "average", "high" and "reliable".

Also it is claimed that monitoring of technical condition of machines and mechanisms carried out by the developer of the equipment. However, as a rule, developers and manufacturers do not develop the system of diagnostics (hydroelectric units, turbine units, the main pumps, rolling mills, etc. are the examples). The systems of diagnostics, as a rule, are developed by specialized firms since specific knowledge and experience are needed for this purpose.

The most critical elements of technical diagnostics are the reliable diagnostic signs which are not to be affected by operating conditions and modes. Currently, as shown by many years of practice, vibrodiagnostics cannot fully solve this problem [6-9]. The vibro-acoustic signal contains the natural vibrational background of the device and information on the interaction of the parts of the operating mechanism. A high level of interference and relatively small changes in the measured signal caused by defects significantly complicate the defects detection. Besides, using the vibrodiagnostics parameters (for example, vibration speed and vibration acceleration), it is not possible to establish a reliable relationship between the measurement results and the mathematical models of machine structure elements and parts and defects to be detected.

So, the problems of traditional diagnostics methods are:

- the lack of automatic and automated systems for assessing the machines and mechanisms current technical state,
- predominance of qualitative evaluations,
- shortening of time between repair intervals,
- big unproductive maintenance costs,
- complexity of detecting nascent defects and providing emergency protection,
- reducing the competitiveness of products.

For the processing of measurement information and decision-making in machines and mechanisms state monitoring, various approaches are currently used: methods of system analysis and synthesis, the theory of pattern recognition, the theory of neural networks, Bayesian methodology, probability theory, reliability theory, wavelet transformations, methods of game theory, methods of artificial intelligence, theory of optimization and decision making, fuzzy sets, fuzzy logic, etc. This wide variety of methods confirms the lack of a unified approach to obtaining information for functioning machines and mechanisms technical state diagnostics.

Thus, in connection with the reasons considered, the traditional methods fail to fully solve the problem of obtaining reliable information on the diagnostic features of cyclic machines and mechanisms for reliable assessment of the current technical state, emergency protection and degradation degree evaluation.

2. Requirements for obtaining effective information.

For obtaining information on functioning of cyclic machines and mechanisms (mechanical and electromechanical systems) characteristic to their technical state it is necessary to meet the following requirements:

1. For assessment of functioning of machines and mechanisms there has to be a possibility of accumulating the measurement information in order to detect the slow trends caused by normal wear and crescent defects. The long observation periods are also necessary to define the diagnostic signs of diagnostic signs defects typical for machines and mechanisms operation.

2. Measuring instruments accuracy is to be sufficient for assessment of the diagnostic signs caused by wear and developing defects which are usually much smaller than the nominal measured parameters values.

3. It is necessary to implement the mathematical models for structural elements, defects and processes happening in the mechanism. It is also necessary to carry out these models identification using experimental data. This increases greatly the diagnostic signs robustness for all types of defects.

3. Phase-chronometric method for diagnosing cyclic machines and mechanisms

The most perspective for cyclic machines and mechanisms are phase methods. Phase approach is most organically connected with a running cycle of the device, and all life cycle of machines and mechanisms is connected with its realization and is supported by all maintenance operations.

Practice of application and research showed that the characteristics received on the basis of phase methods are connected directly to parameters of the mechanism functioning elements and parts[6-12].

The main idea of a phase-chronometric method is registration of kinematic parameters of the running cycle of the parts of the cyclic device. The key controlled parameter of a phase-chronometric method are time intervals corresponding to a running cycle of parts of the device, for example rotation of turbines, shaft of rotor mechanisms. As the running cycle in the normal mode of functioning of a mechanism is characterized by relative stability and repeatability of kinematic parameters at all operational phase, there is a opportunity of detection diagnostic signs corresponding slow processes of wear and growing defects [6-9]. Also, if necessary, measurements are made of the basic operating parameters of the device.

Due to the fact that time intervals are measured with an accuracy unattainable for other physical quantities, phase-chronometric measurements of time intervals variations are highly sensitive to the influence of various factors on the operation of the device, for example:

- accuracies in the manufacturing dimensions of parts and assemblies,
- influence of wear (for example, wear of rolling bearings),
- detection of emerging defects at an early stage, since the phase-chronometric measurements sensitivities is much higher than for vibro-acoustic measurements (because there is a natural background of the machine's vibration),
- the impact of operating conditions,
- the impact of changes in modes of operation, etc.

4. The main advantages of the phase-chronometric diagnostic method

1. Practically achieved high accuracy of measuring time intervals in the conditions of operation of machines up to 1 ± 10^{-7} s provides registration of changes in the operation of the device associated with the interaction of structural elements, processes and operating conditions.

2. The high metrological level makes it possible to obtain individual quantitative characteristics for phase-chronometric diagnostic signs of nascent defects. The metrological level achieved makes it possible to obtain a higher detail of information in comparison with vibrodiagnostics.

3. The most significant advantage of the method is the opportunity of using the mathematical models [6-12] on the basis of classic mechanic theory describing the operation of cyclic mechanical and electromechanical machines and their elements, normal and defected. It enables to evaluate the operation of the machine and identify diagnostic features of defects. The use of mathematical models and algorithms permits to process measurement data and detect defects diagnostic signs.

The mathematical model of the product is the most important factor in the relationship between measurement results and the process of functioning of parts and assemblies in the device [10,11]. In this case, the computational experiment should correspond to the results of processing the experimental data obtained with precision measurements. Comparison of the results of mathematical modeling and experimental data of the functioning of a metal-cutting machine, taken from a rotating spindle in a drive-reduction-spindle system, is presented in [12].

An example of fitting the results of the computational experiment based on mathematical model to experimental data (the intervals of the turnover time and its fractions of the spindle of the machine in the drive-reduction-bearing-spindle system) is shown in the figure 1. The defect of the front bearing of the lathe is presented.

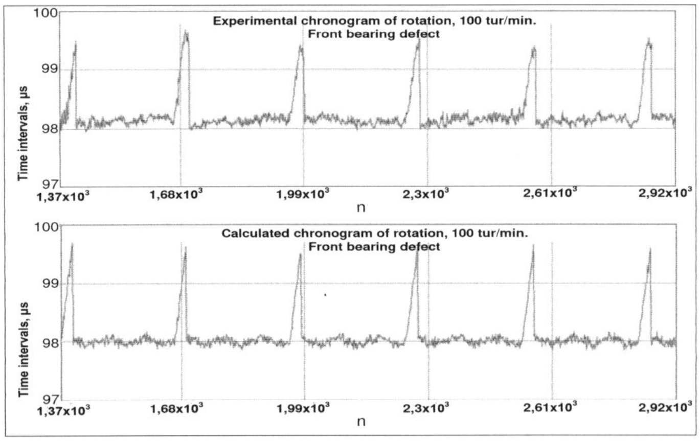

Figure 1. Experimental and settlement chronograms of rotation of a spindle of the lathe with defect of the forward bearing

Thus, the use of the phase method makes it possible to obtain more efficient information and to promote the diagnostics of machines and mechanisms to a new level.

The phase-chronometric method allows performing the analysis of the device operation, defect modeling and defect classification, which minimizes the costs of monitoring, diagnostics and emergency protection of machines and mechanisms.

References

[1] GOST R ISO17359-2015 2016 *Control of a state and diagnostics of machines. General management* (Moscow: Standartinform) (in Russian)

[2] GOST R ISO 13379-2009. 2010 *Control of a state and diagnostics of machines. Guide to interpretation of data and methods of diagnosing* (Moscow: Standartinform) 24 p (in Russian)

[3] GOST 30848-2003 (ISO 13380:2002) 2005 *Diagnosing of Machines on performance data. General provisions* (Moscow: Standartinform) 20 p (in Russian)

[4] GOST R ISO 13373-1-2009 2010 *Control of a State and Diagnostics of Machines.* Vibration *control of a condition of machines. Part 1. General methods* (Moscow: Standartinform) 36 p (in Russian)

[5] GOST ISO 7919-1-2002 2007 *Vibration. Control of a condition of machines by results of measurements of vibration on the rotating shaft General requirements* (Moscow: Standartinform) 28 p. (in Russian)

[6] Tumakova E and Komshin A 2017 *MATEC Web Conf.* **129** 01047

[7] Metelkina E and Komshin A 2017 *MATEC Web Conf.* **129** 01048

[8] Syritsky A 2017 *MATEC Web Conf.* **129** 01049

[9] Kiselyov M I, Komshin A S, Baykov A I, Pronyakin V I and Rudenko A L 2015 *Hydrotechnical construction* **2** 2 (in Russian)

[10] Kiselev M I, Komshin A S, Pronyakin V I, Korsun O N and Nemichev M.Yu. 2011 *Measurement Techniques* **54** 1081

[11] Komshin A S 2013 *Measurement Techniques* **56** 12

[12] Potapov KG 2014 *World of measurements* **12** 10 (in Russian)

IOP Conf. Series: Materials Science and Engineering **312** (2018) 012013 doi:10.1088/1757-899X/312/1/012013

Polymer film strain gauges for measuring large elongations

A P Kondratov[1], A M Zueva[1], R S Varakin[1], I P Taraneç[1] and I A Savenkova[2]

[1]Moscow Polytechnic University, 107023 Moscow, Russia
[2]National Research University Higher School of Economics, 101000 Moscow, Russia

E-mail: apkrezerv@mail.ru

Abstract. The paper shows the possibility to print polymer strain gages, microstrip lines, coplanar waveguides, and other prints for avionics using printing technology and equipment. The methods of screen and inkjet printing have been complemented by three new operations of preparing print films for application of an electrically conductive ink layer. Such additional operations make it possible to enhance the conductive ink layer adhesion to the film and to manufacture strain gages for measuring large elongations.

1. Introduction

Films and fibrous sheets with integrated electrical functions allow creating smart materials and structures with a variety of applications in aircraft engineering and avionics [1], special cloths for test pilots and aerodrome equipment operation safety systems. In the last decade, there has been an increased interest in printed electronics products: strain gages, microstrip lines and coplanar waveguides for microwave devices produced using roll-to-roll planar technology. [2] The most widely used methods are screen and inkjet printing methods using conductive ink and inks [2]. The inkjet method is a printing technology that has the advantage of productivity and variability, i.e. the possibility of a prompt change in the strain gauge shape and size. Most of the conductive inkjet inks contain water as a solvent [2]. Water features a high surface tension and viscosity, which does not allow for forming drops of the optimum size (30–100 μm) by the inkjet printing method [4]. For this reason, the task of developing alternative ways of applying ink and images on flexible polymer substrates, providing an optimal thickness of the conductive layer, remains very relevant. Other important tasks in the strain gage production are increasing interlayer adhesion, enhancing printed layer and substrate resistance to multiple cyclic deformations of large magnitude.

One way to increase the interlayer adhesion in printed images is shown in [5]the paper, that studies the process of applying insulating waterborne ink on the surface of a film resistor printed on a propylene film using the screen printing method. There is an experimental evidence for the effectiveness of screen printing method when applying ink on a "wet layer" of a conductive graphite-containing composition. The graphite-containing composition adhesion to the propylene substrate and waterborne ink adhesion to the conductive layer of a film resistor are rather low; therefore it is not advisable to use this method to manufacture strain gauges and large deformation sensors, resistant to cyclic stretching.

2. Main result

To manufacture strain gages capable of multiple significant stretching without breaking the interlayer adhesion to the polymeric substrate, it is possible to use the "dry crazing" effect of polymers in a state of rigid elasticity [6].

The crystal structure of rigid-elastic films becomes microporous and reversibly reconstructible when stretching and shrinking without plastic deformation. The film elastic deformation reversibility is manifested as a cyclic change in the film color in polarized light [7]. The film microporosity with reversible deformation reaches a maximum at a relative elongation of 20%.

The new way of printing strain gages on polymer films includes three additional operations to increase the adhesion of the conductive ink layer to the elastic substrate. The first operation is stretching and shrinking of the polymer film in a state of rigid elasticity in a gaseous environment (atmosphere). This forms an open pore structure. The second operation is re-stretching the film and fixing it in the stress-strain state in the printing press under a screen form. The third operation is an instant film shrinking after printing an ink layer. This stage involves the "capture" of the liquid ink by closing the substrate pores.

Electrical and mechanical features of strain gages produced using the new method are shown in Fig.1:

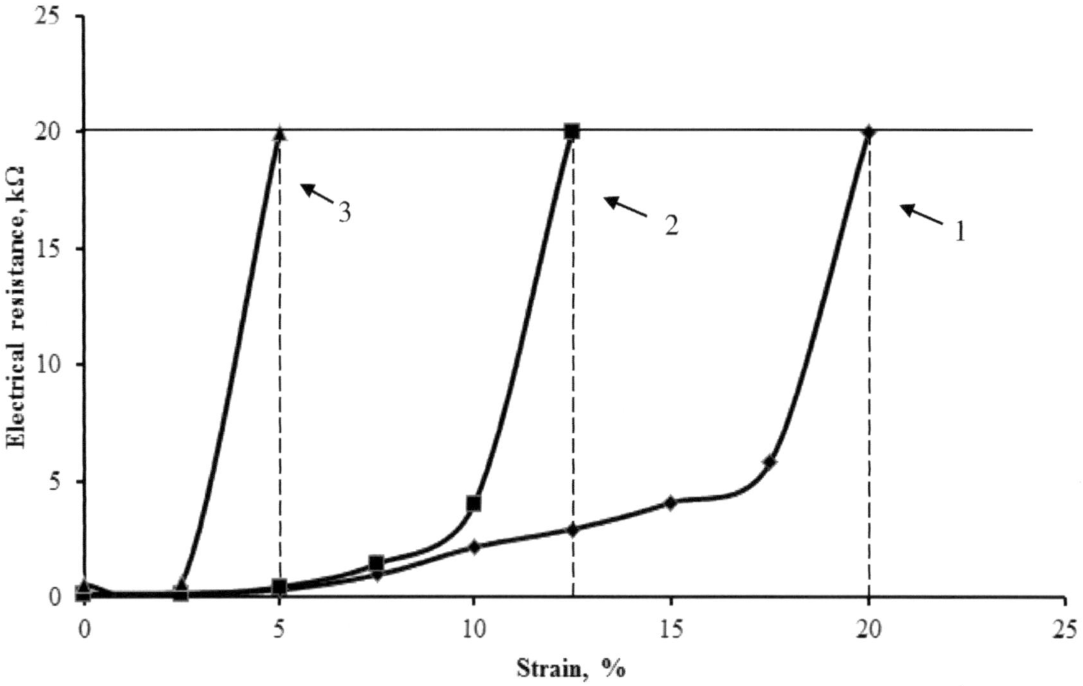

Figure 1. Resistance of strain gages prints obtained by screen printing. The thickness of the conductive ink layers (SunChemical Conductive Graphite 26-8203): 10 (1), 60 (2), 85 (3) μm.

The strain gauge electrical resistance depends on the ink layer thickness. The layer thickness is adjusted by choosing a printing form (screen line number) within a range of 10 to 85 μm. When the ink layer thickness is 10 μm, the linear strain measurement range with a strain sensitivity of 600 is 0 to 18%. When the ink layer thickness is 60 μm, the linear strain measurement range with a strain sensitivity of 1600 is 0 to 8%. When the ink layer thickness is 85 μm, the linear strain measurement range with a strain sensitivity of 700 is 0 to 3%. Strain gage prints with different ink layer thickness can be used as mechanical switches or open circuit sensors in electrical schemes, which are programmed to a certain resistance by choosing the ink layer thickness. The threshold resistance of 20 kΩ with different layer thicknesses is achieved at 5, 12, 20 % elongation. The new printing method allows for manufacturing film strain gages and elastic strain gages of up to 70% and more. For this purpose, it is necessary to change the method of applying the conductive ink on a stress-strain film and to use conductive composition of lower viscosity.

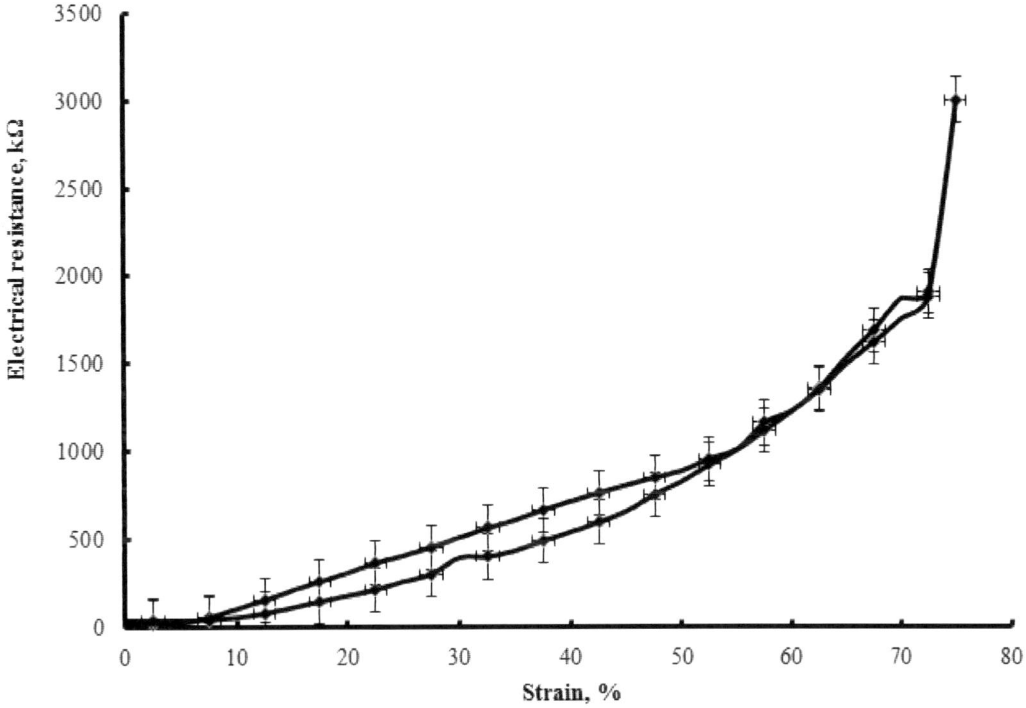

Figure 2. Resistivity of fingerprints of strain gages obtained by spraying a graphite spray (model Kontakt Chemie GRAPHIT 33) onto a film of rigid polypropylene at the time of extension by 70%. The thickness of the layer of electrically conductive ink is 5 μm.

Fig. 2 shows the strain sensitivity of the strain gage print manufactured by spraying a conductive composition spray onto a rigid polypropylene film using the suggested printing method. The measuring range of the relative tensile strain is 0 to 70%. The electrical resistivity dependence on the relative elongation is described with high reliability (R^2=0.95) by the following equation:

$$\Delta Q/Q_o = G_* \varepsilon^2 \text{ or, approximately, } (R^2=0{,}87) \quad \Delta Q/Q_o \approx G_* \varepsilon,$$

Where: **Q** is the strain gage electrical resistance (Ohm), **G** is strain sensitivity and **ε** is relative extension (%).

3. Conclusion

Different strain gauge resistance values at the moments of stretching and shrinking do not exceed the measurement error, provided that the relative deformation values are equal (40%). Resistance of these strain gage parameters to large cyclic deformations is due to the features of the polymer substrate structure and the mechanism for fixing the ink layer on the surface of a rigid polypropylene film. The polypropylene film deforms to form a plurality of nano- and micropore crazes [8], the highly developed surface of which is formed during the first operation of stretching the printed substrate in an air environment. The high specific energy of the new craze surface is lower due to adsorption of fluid molecules and ink ingredients. The instant film shrinking after it's contact with liquid ink allows for mechanically capturing the conductive dye particles and moving the same to the pore volume. After removing solvents and craze structure collapse upon reducing the film length [9], particles of the conductive dye, concentrated on its surfaces inside the film, come into contact and, due to multiple

IOP Conf. Series: Materials Science and Engineering **312** (2018) 012013 doi:10.1088/1757-899X/312/1/012013

contacts with each other, ensure a high electrical conductivity of a strain gage at large cyclic deformations.

References

[1] Mishra A and Mishra D 2010 *An Exploratory Review. Promet – Traffic&Transportation* **22**(5) 363.

[2] Stoppa M and Chiolerio A 2014 *Sensors* **14(7)** 11957.

[3] Khairilhijra K, Tan S C and Khairul A M 2016 *Journal of Engineering and Applied Sciences* **11**(10) 6619.

[4] Huang D and Liao F 2003 *Journal of the Electrochemical Society*, **150**(7) 412-417.

[5] Kondratov A P, Zueva A M and Nagornova I V 2017 Parameters dynamics estimation method for printed electronics conductive elements layers *2017 IEEE Dynamics of Systems, Mechanisms and Machines (Dynamics) (Omsk, Russia)*

[6] Kondratov A P 2014 *Light & Engineering Svetotekhnik* **22**(3) 74.

[7] Varepo L G, Ermakova I N, Nagornova I V and Kondratov A P 2017 *AIP Conference Proceedings* **1876** 020080

[8] Rukhlya E G, Yarysheva L M, Volynskii A L and Bakeev N F 2016 *Physical Chemistry Chemical Physics* **18**(14) 9396

[9] Trofimchuk E S, Efimov A V, Nikonorova N I, Volynskii A L, Bakeev N F, Nikitin L N, Khokhlov A R and Ozerina L A 2017 *Polymer science. Series A* **53**(7) 546

IOP Conf. Series: Materials Science and Engineering **312** (2018) 012014 doi:10.1088/1757-899X/312/1/012014

Optimal pattern synthesis for speech recognition based on principal component analysis

O N Korsun[1] * and A V Poliyev[2]

[1]State Research Institute of Aviation Systems, Moscow, 125319, Russia
[2]Moscow Institute of Physics and Technology, Moscow, 117303, Russia

E-mail: marmottoduo@yandex.ru

Abstract. The algorithm for building an optimal pattern for the purpose of automatic speech recognition, which increases the probability of correct recognition, is developed and presented in this work. The optimal pattern forming is based on the decomposition of an initial pattern to principal components, which enables to reduce the dimension of multi-parameter optimization problem. At the next step the training samples are introduced and the optimal estimates for principal components decomposition coefficients are obtained by a numeric parameter optimization algorithm. Finally, we consider the experiment results that show the improvement in speech recognition introduced by the proposed optimization algorithm.

1. Introduction

A reliable and rational man-machine interface is an important task of modern engineering [1-3]. In recent years in avionics we witness a considerable growth of attention to audio interfaces and speech recognition [4-10], since their successful implementation may improve the flight safety and reduce the pilots' workload [1,10]. For example, on Eurofighter Typhoon since 2005 a speaker-dependent system based on the comparing with the patterns is employed.

The principal requirement for speech interface in avionics is high probability of correct word recognition even under strong noises, because an error in aircraft control may be critical for the flight safety. In addition, the time complexity of the algorithm is an important point for onboard software.

For the recognition of speech commands, the method of comparison with a pattern shows a good performance for speech control of aviation equipment. In this work, one proposes the new method for a pattern generation and its improvements using the principal component method. The results of word recognition with new patterns are presented in the experimental part. And, in conclusion, one considers a way to simplify the algorithm in order to reduce the working time while preserving the results.

2. Speech signal parametric portrait in automatic speech recognition

The traditional methods for automatic recognition of speech commands are based on the time-spectral quantisation of an input word record. The transform used in this article obtains a time-spectral parametric portrait and includes the following steps. In the beginning, a speech signal is divided into equal intervals of 10-30 ms. After that, one uses the standard digital signal processing procedures, such as amplification of high-frequency signal components, interval weighing by Hannah window, fast Fourier transform, averaging over frequencies and logarithm of spectral densities [11]. Each of these intervals is projected into 30-40 frequency bands [1]. As a result, for each of the time intervals bands we obtain the estimates of spectral densities logarithms in 30-40 frequency bands [4,8,10]

These parameters, presented in a form of a matrix, we call the parametric portrait of the word. The columns of this portrait characterize the spectral decomposition of the speech signal for each time interval.

Further, the pattern parametric portrait is compiled from available parametric portraits of the speech commands records. The simplest way to obtain a pattern is to use the mean of the parametric portraits of several similar records. Then, the parametric portrait of the recognized word is compared with the pattern parametric portrait. This comparison may be carried out by the criterion of the maximum correlation between the elements of parametric portrait matrices. After comparing the recognized word with all patterns, the pattern that provides the correlation maximum is selected. The speech command associated to this pattern is recognized as the result of automatic recognition.

3. Principal component analysis

The principal component analysis (PCA) is traditionally used to reduce the number of dimension of the observed vectors space without significant loss of information. The PCA method applications cover a great number of fields from statistics to medicine [12]. The principal components are the orthogonal vectors that form the orthogonal coordinates system in a space of observed data.

Let us consider an initial set of vectors X in a linear space $\boldsymbol{L}^p$. The principal component analysis enables to switch to a basis $\boldsymbol{L}^{p'}, (p' \ll p)$ such that the first component (the first vector of a basis) corresponds to the direction along which the variance of the initial set of vectors is maximal. The direction of the second component (the second vector of a basis) is selected so that the variance of the original vectors along it is maximal provided that it is orthogonal to the first vector of a basis. The remaining basis vectors are defined similarly. As a result, the directions of the basis vectors, that are selected to maximize variance of the initial set along the first components, are called the principal components. It turns out that the main variability of the initial set of vectors is presented in the first few components, and it is possible to move to the space of smaller dimension discarding the remaining (less significant) components [12].

Let us consider a multi-dimensional random observation $X = \begin{pmatrix} x^{(1)} \\ \vdots \\ x^{(p)} \end{pmatrix}$. The problem is to reduce the dimension of X from p to p' with no considerable loss of information. One can solve this problem by defining all possible linear combinations of orthogonal normalized combinations of initial observations $z^{(j)}(X) = c_{j1}(x^{(1)} - \mu^{(1)}) + \cdots + c_{jp}(x^{(p)} - \mu^{(p)})$, where $[\mu^{(1)}, \dots, \mu^{(p)}]'$ - vector of observation X means.

A measure of information value for p'-dimensional system of $(z^{(1)}(X), \dots, z^{(p')}(X))$ may be defined as $I_{p'} = \frac{Dz^1 + \cdots + Dz^{p'}}{Dx^1 + \cdots + Dx^p}$, where Dz is the operation calculating the random variable variance. It can be shown that all p principal components of X may be represented as $Z = LX$, where $Z = (z^{(1)}, \dots, z^{(p)})', X = (x^{(1)}, \dots, x^{(p)})'$, and matrix L consists of rows $l_j = (l_{j1}, \dots, l_{jp}), j = \overline{1,p}$, which are the eigenvectors of the random variable X covariance matrix, $\sum = \sigma_{ij}, i, j = \overline{1,p}$. The components are enumerated according to decrease of corresponding eigenvalues $\lambda_j, j = \overline{1,p}$.

The basic properties of principal components:

- L is an orthogonal matrix, that is $LL' = L'L = I$, where I – identity matrix;
- The covariance matrix of the principal components vectors $\sum_Z = L\sum L' = \begin{pmatrix} \lambda_1 & \cdots & 0 \\ \vdots & \ddots & \vdots \\ 0 & \cdots & \lambda_p \end{pmatrix}$;
- The sum of observation components $x^{(1)}, \dots, x^{(p)}$ variances is equal to the sum of the all principal components $z^{(1)}, \dots, z^{(p)}$ variances;
- Criterion of PCA information value is $I_{p'}(Z(X)) = \frac{Dz^1 + \cdots + Dz^{p'}}{Dx^1 + \cdots + Dx^p} = \frac{\lambda_1 + \lambda_2 + \cdots + \lambda_{p'}}{\lambda_1 + \lambda_2 + \cdots + \lambda_p}$, where $\boldsymbol{\lambda}_1, \boldsymbol{\lambda}_2, \dots, \boldsymbol{\lambda}_p$ – the eigenvalues of the covariance matrix $\sum$ of the vector X arranged in decreasing order. One can use this criterion as a reference point of view to make decide about the principal components to be withdrawn with no considerable damage to the information content, thereby reducing the dimension of the space used.

4. Development and validation of the algorithms for spectral decomposition of a word portrait on principal components

Let us consider parametric portraits of M different samples of the same word $\{x_{lj}(k)\}, k = 1, 2, \dots, M; l = 1, 2, \dots N_t; j = 1, 2, \dots, N_f$. Let us transform a matrix portrait for each k into a one-dimensional array with a number of elements equal to $P = N_f N_t$. Therefore, we have M vectors each with P dimensions $\{x_{ik}\}, k = 1, 2, \dots, M; i = 1, 2, \dots, P$. Explicitly:

$$x_1 = \begin{bmatrix} x_{11} \\ x_{21} \\ \dots \\ x_{p1} \end{bmatrix}, x_2 = \begin{bmatrix} x_{12} \\ x_{22} \\ \dots \\ x_{p2} \end{bmatrix}, \dots, x_M = \begin{bmatrix} x_{1M} \\ x_{2M} \\ \dots \\ x_{pM} \end{bmatrix} \quad (1)$$

Let us assemble these M vectors into a matrix of dimension $p \times M$:

$$X = [x_1 x_2 \dots x_M] = \begin{bmatrix} x_{11}\ x_{12}\ \dots x_{1M} \\ x_{21}\ x_{22}\ \dots x_{2M} \\ \dots\ \dots\ \dots\ \dots \\ x_{p1}\ x_{p2}\ \dots x_{pM} \end{bmatrix} \quad (2)$$

Further we may estimate the matrix of correlation moments for vectors $x_1, x_2, \dots, x_M$ of dimension M

$$K_x = X^T X = \begin{bmatrix} x_1^T x_1\ x_1^T x_2\ \dots x_1^T x_M \\ x_2^T x_1\ x_1^T x_2\ \dots x_2^T x_M \\ \dots\ \dots\ \dots\ \dots \\ x_M^T x_1\ x_M^T x_2\ \dots x_M^T x_M \end{bmatrix} \quad (3)$$

One may note that each element of the matrix K_x is a scalar product of the corresponding vectors and so matrix K_x is symmetrical. Therefore, it is possible to calculate M eigenvalues $\lambda_1, \lambda_2, \dots, \lambda_M$ and corresponding M-dimensional eigenvectors $l_1, l_2, \dots, l_M$ for this matrix. The eigenvalues may be arranged in the descending order $\lambda_1 > \lambda_2 > \dots > \lambda_M$.

The first principal component z_1 is defined as a linear combination of the original vectors $x_1, x_2, \dots, x_M$ with coefficients equal to the elements of the first eigenvector:

$$l_1^T = [l_{11} l_{21} \dots l_{M1}]: z_1 = l_{11} x_1 + l_{21} x_2 + \dots + l_{M1} x_M = \sum_{i=1}^{M} l_{i1} x_i \quad (4)$$

Similarly, the principal components $j = 2, 3, \dots, M$ are calculated using the formula:

$$z_j = \sum_{i=1}^{M} l_{ij} x_i \quad (5)$$

According to the theory described above the total principal components variance is equal to the total variance of the initial vectors, i.e. transform does not alter the energy of the signal. The variance of the principal components z_j is equal to the corresponding eigenvalue λ_j. So the behavior of the system is determined mainly by the first few principal components z_j, $j = 1, 2, \dots, M'$. It allows us to reduce the dimension of the problem and consider M' principal components instead of the M initial vectors.

To estimate the errors introduced during the transition from the original vectors M to M' principal components one can use the expression:

$$I_p = \frac{\lambda_1 + \lambda_2 + \dots + \lambda_{M'}}{\lambda_1 + \lambda_2 + \dots + \lambda_M}. \quad (6)$$

It is clear that the information value criterion (6) is equal to the ratio of variance, explained by the M' selected components, to the total variance of the initial signal.

5. Development of the algorithm for optimal patterns synthesis based on principal components analysis

In this work, the recognition is performed for three Russian words: "pilotazh", "masshtab" and "navigatsiya", which are pronounced by ten different speakers without noise and by four speakers with noise in the headphones.

The noise in the headphones is implemented as follows. The speaker puts the headphones on, which play the noise of the given volume. Also the speaker can hardly hear his own voice, i.e. there is no feedback in the headphones. The noise is not recorded along with the spoken word, and is only played in the headphones. This procedure is performed not to analyze the influence of noise on the recognition process, but to evaluate the effect of noise on changes of the speaker's speech.

The normal patterns for the recognition of each of the three words were formed as mean parametric portraits of ten speakers with no noise. It is known that the noise in the headphones changes the speech, so the normal patterns formed in no noise conditions are not fully adequate to the records of speakers with noise in the headphones. The idea of the experiment was to adjust normal patterns to the data with noise in the headphones using principal components, numeric optimization and additional training samples.

The principal components were formed as follows. Speech samples of ten speakers with no noise were introduced to build ten separate pattern, one pattern vector for each speaker. The parametric portraits of these $M = 10$ patterns were considered further as original vectors $x_1, x_2, \dots, x_M$ (1). Then according to (2)-(5) $M = 10$ principal components were built: z_j , $j = 1,2, \dots, M$. Then the formula (6) was applied which enabled to select $M' = 6$ components at information value criterion (6) equal to 0.95. At the next step, we expressed the normal patterns as a linear combination of selected principal components:

$$e_{pat\_norm} = k_0 Z_0 + k_1 Z_1 + \dots + k_6 Z_6 . \qquad (7)$$

The coefficients $k_0, \dots, k_6$ were obtained through the multiple linear regression; the coefficient k_0 was added to incorporate the mean of normal pattern. Naturally, the decomposition (7) was applied separately to each of normal patterns that is the patterns for three words in consideration. Now we should note that usually a parametric portrait of a word (recognition pattern is also a parametric portrait) is a vector that includes 1200-1800 elements. So the PCA enabled us to reduce the set of parameters to a few coefficients $k_0, \dots, k_6$.

For normal pattern optimization it is necessary to form an optimization criterion and a training sample. The natural criterion for a search of coefficients $k_0, \dots, k_6$ is the minimum of recognition errors: $N_{errors} \to min$. The fault is that the criterion values are discreet. The problem of three words recognition is quite simple, so the number of errors at training sample may be inconsiderable or even equal to zero. As a result, the discrete criterion may not be sensitive enough. In this research we used the continuous scalar measure of recognition quality ΔF. This measure was introduced in [13]. For each recognized word the measure ΔF is equal to Fisher z-transforms difference between maximum correlation and correlation nearest to maximum. That is, the greater is distinction between the «winner» word and «the best of the rest» words, the higher is the recognition quality.

For training sample we selected one out of four speaker records with noise in the headphones. More precisely, we chose N-v speaker records with the 80 dB noise in the headphones. The search of coefficients $k_0, \dots, k_6$ was performed by a widely known coordinate descent method. To achieve an extreme point 10-15 iterations proved to be enough. The optimization algorithm was applied to the patterns of all the three words, so the optimization problem dimension was equal to 21. For the starting values of the optimized parameters we assumed the coefficients of normal patterns linear decomposition (7).

6. Results

To estimate the optimized pattern efficiency we performed the automatic recognition tests. At this step we used all four speakers records with two levels of noise in the headphones: 80 dB and 90 dB, so the number of testing sets was equal to eight. The results of the optimized pattern application have shown the improvement in samples recognition not only for the N-v words, but also for other speakers outside the training set. The total number of errors before optimization for the three words was equal to 3.15% and after optimization has reduced to 0.89%. The results are shown in Tables 1 and 2.

Table 1. The samples recognition results with the 80 dB noise in the headphones with normal (1) and optimized (2) patterns.

Speaker	Word	Errors 1	Errors 2
B-k	pilotazh	6	0
	masshtab	0	0
	navigatsiya	0	8
G-v	pilotazh	3	0
	masshtab	0	0
	navigatsiya	0	0
N-v	pilotazh	8	0
	masshtab	0	0
	navigatsiya	0	0
F-v	pilotazh	9	0
	masshtab	0	0
	navigatsiya	0	0

Table 2. The samples recognition results with the 90 dB noise in the headphones with normal (1) and optimized (2) patterns.

Speaker	Word	Errors 1	Errors 2
B-k	pilotazh	7	0
	masshtab	0	0
	navigatsiya	0	1
G-v	pilotazh	3	0
	masshtab	0	0
	navigatsiya	0	2
N-v	pilotazh	12	0
	masshtab	0	0
	navigatsiya	0	0
F-v	pilotazh	4	0
	masshtab	0	0
	navigatsiya	8	4

7. Conclusion

The patterns, obtained by optimizing the principal components coefficients, showed significantly fewer errors in the recognition for the majority of samples. This method proved to be the an effective way to improve word recognition results, along with the words splitting into homogeneous parts [10], which also gives improved results for this problem. The achieved positive results confirm validity of proposed the method, which includes PCA based reduction of parameter space dimension and pattern optimization. The further progress is to be obtained by solving recognition problems with greater vocabulary using more sophisticated optimization algorithms [14]

Acknoledgements
The research is supported by Russian fund for basic research (RFBR), project 15-08-06946-a.

References

[1] Evdokimenkov V.N., Kim R.V., Krasil'shchikov M.N., Sebryakov G.G. 2015 *Journal of Computer and Systems Sciences International*. 54(4) 609
[2] Polyak B T and Khlebnikov M V 2017 *Automation and Remote Control* 3 130
[3] Kolokolov A.S. 2006 *Problemy upravleniya* 3 13–18 (In Russian)
[4] Rabiner L.R., Juang B.-H. 1993 *Fundamentals of Speech Recognition* (Englewood Cliffs. NJ: Prentice-Hall)
[5] Springer Handbook on Speech Processing and Speech Communication 2007 (Berlin, Heidelberg: Springer)
[6] Benesty J, Sondhi M M, Huang Y A 2008 *Springer Handbook of Speech Processing* (Berlin: Springer)
[7] Schmidt-Nielsen A, Marsh E, Tardeli J, Gatewood P, Kreamer E, Tremain T, Cieri C and Wright J 2000 *Speech in Noisy Environments (SPINE) Evaluation Audio* (Philadelphia, PA: Linguistic Data Consortium)
[8] Kolokolov A.S., Lublinsky I.A. 2015 *Automation and Remote Control* 10 144–151
[9] Savchenko L V 2014 *Informatsionno-upravlyayushchiye sistemy* 1 23 (In Russian)
[10] Korsun O N and Poliev A V 2016 *Journal of Computer and Systems Sciences International* **55**(4) 609
[11] Oppenheim A V and Shafer R V 2010 *Discrete-Time Signal Processing* (Englewood Cliffs. NJ: Prentice-Hall)
[12] Ayvazyan S A, Bukhtshtaber V M, Yenyukov I S and Meshalkin L D 1989 *Prikladnaya statistika: Klassifikatsiya i snizheniye razmernosti* 607 (Moscow: Finansy i Statistika) (In Russian)
[13] Korsun O N and Gabdrakhmanov A S 2017 *Vestn. komp'yuternykh i informatsionnykh tekhnologiy* 1 10-15(In Russian)
[14] Luenberger D G, Yinyu Y 1984 *Linear and nonlinear programming* vol 2 (Reading, MA: Addison-wesley)

IOP Conf. Series: Materials Science and Engineering **312** (2018) 012015 doi:10.1088/1757-899X/312/1/012015

Aircraft aerodynamic coefficients identification using flight tests data

O N Korsun[1]*, A V Stulovskii[1], O A Balyk[2] and M V Zolotaeva[3]

[1]State Research Institute of Aviation Systems, Moscow, 125319, Russia
[2] Flight Dynamics and Control Department, Moscow Aviation Institute (National Research University), Moscow, 125993, Russia
[3]Gubkin Russian State University of Oil and Gas (National Research University), Moscow, Russia

* E-mail: marmotto@rambler.ru

Abstract. This paper discusses the problems of aerodynamic coefficients identification from flight test data. The principal stages of identification procedure are considered, the more sophisticated stages such as data compatibility check and parameter estimation from flight data are discussed in details. The paper also presents the examples of flight test data processing, which confirm the opportunities for on-board measurement systematic error detection and wind tunnel aerodynamic parameter values correction.

1. Introduction

The system identification methods form an important part of flight test data analysis. Through these methods aircraft parameter estimates may be obtained [1-4], which is important for numerous applications, such as simulators and flight control systems design, aircraft modernization etc. Examples for some applications may be found in [5-7].

Technology of aircraft's mathematical model parameter identification is presented in [1-4]. It includes the following basic steps[1-4]:

- collecting a priory data of the aircraft, aerodynamic parameters, engine, control system from aircraft developer company;
- formulating the requirements for the on-board registration system, above all the list of registered parameters, the acceptable levels of measurements errors and registration frequency;
- analysis of a priory data and flight experiment design;
- running the flight test in order to gather experimental data for identification;
- checking the compatibility of aircraft's parameters recorded by on-board equipment;
- processing of the flight tests results using the identification methods and algorithms;
- comparing the identification estimates with a priory aerodynamic data.

2. Data compatibility check

Checking the recorded flight data compatibility and identification of lateral and longitudinal aerodynamic coefficients are the most complicated stages. So let us discuss them in details. Checking the compatibility is achieved by comparing the signals recorded during the flight with the results

IOP Conf. Series: Materials Science and Engineering **312** (2018) 012015 doi:10.1088/1757-899X/312/1/012015

obtained by numerical solution of the aircraft dynamic differential equations [1, 4]. Let us examine the examples of processing the flight test data.

Figures 1 and 2 show the results of the data compatibility check. Figure 1 represents the fitting of pitch and roll signals in case of normal functioning of the on-board measuring system. The figure shows that when measurements are correct the plots of measured and simulated pitches and rolls are visually indiscriminate.

Figure 1. The pitch (plotted against left ordinate axis, degrees) and roll signals(plotted against right ordinate axis, degrees) fitting simulated (TETM(L) and GAMM(R)) and measured in flight (TET(L) and GAM(R)) at normal work of measuring and recording system.

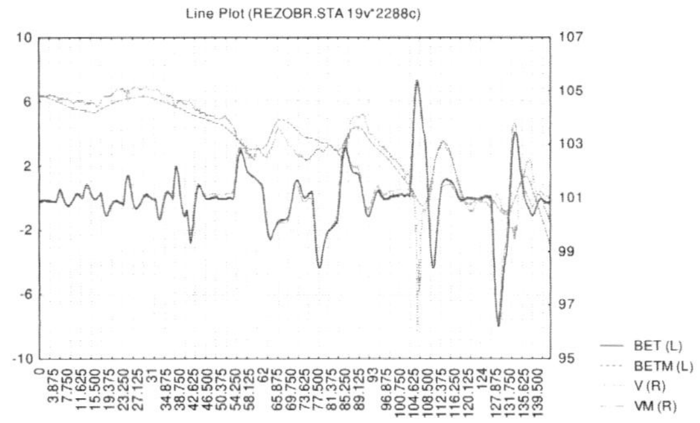

Figure 2. The error of barometric measurement of airspeed(measured airspeed V(R) deviates abruptly from simulated airspeed VM(R), plotted against the right ordinate axis, m/s), caused by sharp change of sideslip angle (measured sideslip BET(L) and simulated sideslip BETM(L) plotted against the left ordinate axis, degrees, coincide with high accuracy).

Figure 2 shows the opportunities for error detection. It shows that measured and simulated signals of sideslip angle and airspeed normally coincide with high accuracy, whereas at a sharp change of sideslip angle by about 9 degrees the measured airspeed deviates by 5 m/s from the simulated value. It is evident that in case of airspeed barometric sensor, positioned on the fuselage surface, error increases with the growth of sideslip angle. This error is clearly systematic and is to be corrected.

The experience shows that the data compatibility check is capable of detecting systematic measurement errors for such major flight parameters as angles of attack and sideslip, overloads, altitude and airspeed, etc. This approach may also detect the effect of the coordinates of the sensor position, and time shifts caused by insufficient synchronization of heterogeneous data flows registered on-board. Identification of the measurement errors caused by the wind is discussed in [8].

3. Aircraft aerodynamic coefficients identification

It is reasonable to carry out identification separately for longitudinal and lateral movement. The corresponding equations are selected from the general nonlinear dynamic model of the aircraft spatial movement [9]. The object model takes the form:

longitudinal channel

$$\frac{d\alpha}{dt} = \omega_z - \frac{1}{\cos\beta}\left[\left(\frac{a_x}{V} - \omega_y \sin\beta\right)\sin\alpha + \left(\frac{a_y}{V} + \omega_x \sin\beta\right)\cos\alpha\right],$$
$$\frac{d\omega_z}{dt} = \frac{I_x - I_y}{I_z}\omega_x\omega_y + q\frac{S b_A}{I_z} m_z - \frac{k\omega_y}{I_z} - \frac{(P_{right} + P_{left})y}{I_z}; \qquad (1)$$

lateral channel

$$\frac{d\beta}{dt}=\frac{a_z}{V}\cos\beta-\left(\frac{a_x}{V}\sin\beta-\omega_y\right)\cos\alpha+\left(\frac{a_y}{V}\sin\beta+\omega_x\right)\sin\alpha,$$

$$\frac{d\omega_x}{dt}=\frac{I_y-I_z}{I_x}\omega_y\omega_z+q\frac{S\,l}{I_x}m_x \qquad (2)$$

$$\frac{d\omega_y}{dt}=\frac{I_z-I_x}{I_y}\omega_x\omega_z+q\frac{S\,l}{I_y}m_y+\frac{k\omega_z}{I_y}+\frac{(P_{right}-P_{left})z}{I_y};$$

accelerations along axes of body-fixed coordinate system

$$a_x=\frac{qS(-c_x+c_p)}{m}-g\sin\upsilon,$$

$$a_y=\frac{qSc_y}{m}+g\cos\upsilon\sin\gamma, \qquad (3)$$

$$a_z=\frac{qSc_z}{m}+g\cos\upsilon\sin\gamma;$$

overloads along axes of body-fixed coordinate system:

$$n_x=\frac{qS(-c_x+c_p)}{mg},\; n_y=\frac{qSc_y}{mg},\; n_z=\frac{qSc_z}{mg}, \qquad (4)$$

where α, β – angles of attack and sideslip, rad; ω_x, ω_y, ω_z – angular velocities related to body-fixed coordinate system, rad/s; υ, γ, ψ – angles of pitch, roll, yaw, rad; V – airspeed, m/s; H – flight altitude, m; m_x, m_y, m_z – coefficients of aerodynamic torques; c_x, c_y, c_z – coefficients of aerodynamic force in body-fixed coordinate system; I_x, I_y, I_z, I_{xy} – moments of inertia related to axes of body-fixed coordinate system, kg m^2; m – aircraft weight, kg; l, b_A – wing span and length of mean aerodynamic chord, m; S – wing surface area, m^2; $q = \rho_H V^2/2$ – dynamic air pressure, Pa; ρ_H – air density at the altitude of flight, kg/m^3; $c_p = P/qS$ – thrust coefficient; P_{right}, P_{left} – thrust of right and left engine, N; k – kinematic momentum of engine rotors, kg m^2/s; y, z – engine coordinates in body-fixed coordinate system, m.

For coefficients of aerodynamic torques m_x, m_y, m_z and forces c_x, c_y, c_z linearized and nonlinear decomposition are used.

In case of the identification in the longitudinal channel state or output vector is $\mathbf{y}^T(t)=[\alpha(t)\;\omega_z(t)]$, and scalar input signal is $u(t)=[\varphi_B(t)]$ or $u(t)=[\delta_B(t)]$, where $\varphi_B(t)$ - angle of symmetric stabilizer deflection, $\delta_B(t)$ - angle of elevator deflection.

When identifying the lateral channel the state vector is $\mathbf{y}^T(t)=[\beta(t)\;\omega_x(t)\;\omega_y(t)]$, and input vector $\mathbf{u}^T(t)=[\delta_A(t)\;\delta_R(t)\;\Delta\varphi_{st}(t)]$, where $\Delta\varphi_{st}(t)$ - angle of differential stabilizer deflection, $\delta_A(t)$, $\delta_R(t)$ - angles of aileron and rudder deflection.

In numerical integration of the aircraft dynamics equations we substitute to the right part of equations the measured values of signals which are not included in the state vector. For example, for longitudinal movement these are the signals $\beta(t), \omega_x(t), \omega_y(t), H(t), V(t)$. Values of engine thrust are substituted based on the available engine thrust model. While defining the problem of identification it is necessary to keep in mind that separate identification of the forces of air drag and engine thrust is normally impossible. The traditional solution is to estimate engine thrust, for example, from the available engine model and carry out identification of air drag force, or vice versa [1-4].

Observation models for the identification of the longitudinal and lateral channel coefficients are given by, respectively

$$\mathbf{z}(t_i)=\begin{bmatrix}\alpha(t_i)\\ \omega_z(t_i)\\ n_y(t_i)\end{bmatrix}+\boldsymbol{\eta}_B(t_i), \qquad (5)$$

$$\mathbf{z}(t_i) = \begin{bmatrix} \beta(t_i) \\ \omega_x(t_i) \\ \omega_y(t_i) \\ n_z(t_i) \end{bmatrix} + \boldsymbol{\eta}_b(t_i),$$

where $\boldsymbol{\eta}_B(t_i)$, $\boldsymbol{\eta}_b(t_i)$ - vector measurement noise in longitudinal and lateral channels.

4. Examples

Examples of parameter identification for various aircraft from flight tests data appear below. So, figures 3-4 show roll and yaw angular rates for the a priory model, indicating remarkable error between model and measurements. After identification (interceptors' efficiency proved to be 10% of the a priory value) the error reduces considerably (figures 5-6).

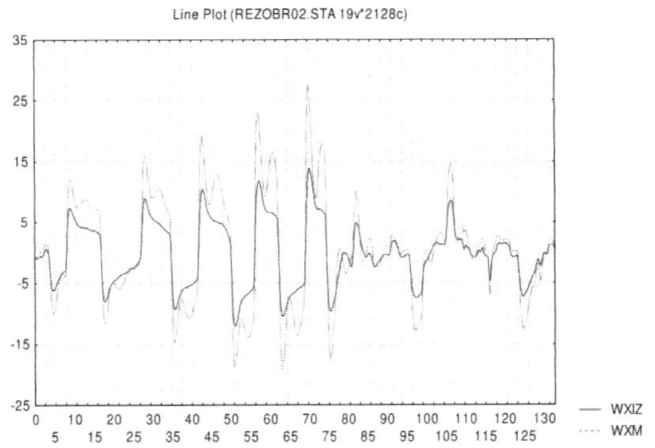

Figure 3. Measured and simulated roll rate.

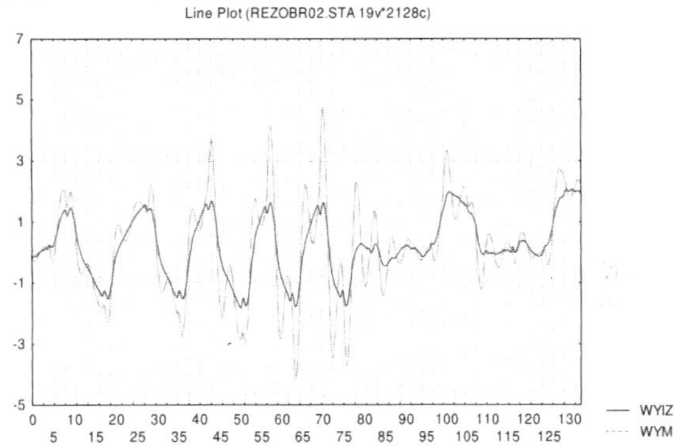

Figure 4. Measured and simulated yaw rate.

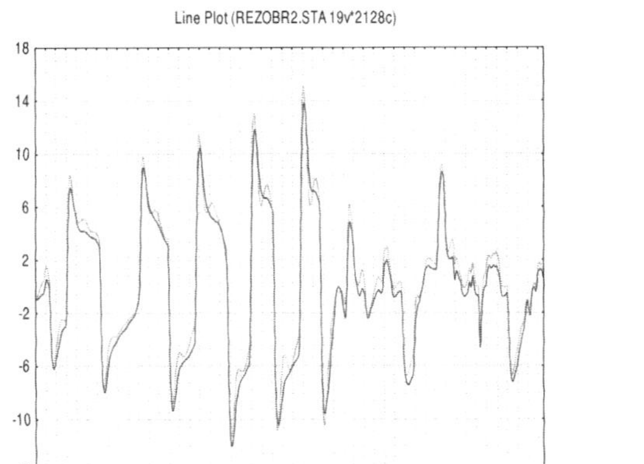

Figure 5. Measured and simulated roll angular rate after adjustment of the aerodynamic data.

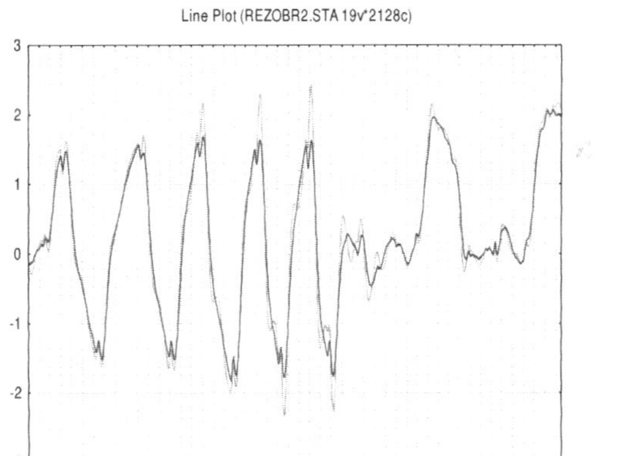

Figure 6. Measured and simulated yaw angular rate after adjustment of the aerodynamic data.

Thus, through identification we have obtained that the interceptors' efficiency is 10...20% of a priory values drawn from wind tunnel experiments. So the technology of identification enables to adjust the initial data bank of aerodynamic coefficients.

Another example of lateral parameter identification appears in figure 7. In this case, coherence of identification estimates and the appropriate coefficients from wind-tunnel aerodynamic data bank is rather high.

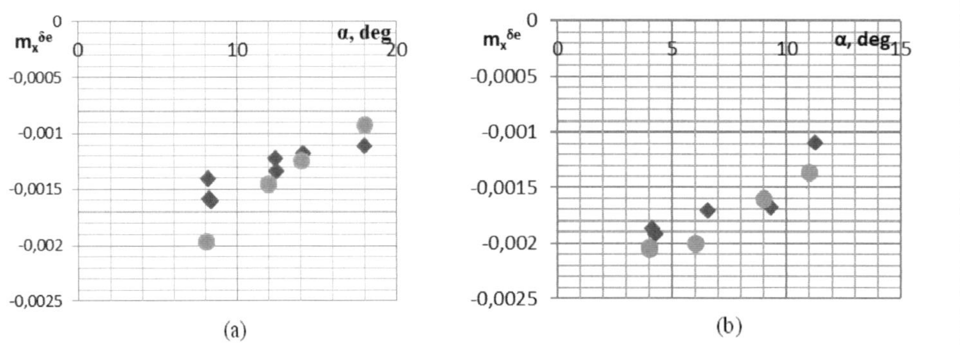

Figure 7. Comparison of aileron efficiency $m_x^{\delta_0}$ estimated by identification (blue) and wind-tunnel aerodynamic data, where (a) the take-off configuration and (b) landing configuration.

Acknowledgments
This work was supported by Russian Foundation for Basic Research (RFBR), project 15-08-06237-a.

References

[1] Klein V 1989 *Progress in Aerospace Sciences* **26** (1) 1–77
[2] Klein V and Morelli E A 2006 *Aircraft System Identification: Theory and Practice* (Reston, VA: AIAA, Inc.)
[3] Jategaonkar R V 2006 *Flight Vehicle System Identification: a Time Domain Methodology* (Reston, VA: AIAA, Inc.)
[4] Korsun O N and Poplavsky B K 2014 Approaches for flight tests aircraft parameter identification, *Proc. of the 29 Congress of Int. Council of the Aeronautical Sciencies* (Saint Petersbourg: ICAS) ICAS 2014_0210
[5] Carlson H A , Verberg R, Hemati M S and Rowley C W 2015 A flight simulator for agile fighter aircraft and nonlinear aerodynamics *53rd AIAA Aerospace Sciences Meeting* (Reston, VA: AIAA SciTech Forum) 2015-1506
[6] Schutte A, Einarsson G, Raichle A, Schoning B, Monnich W and Forkert T 2009 *J of Aircraft* **46**(1) 53-64
[7] Luchtenburg D M, Rowley C M, Lohry M W, Martinelli L and Stengel R F 2015 *J of Aircraft* **52**(3) 890-5
[8] Korsun O N, Nikolaev S V, Pushkov S G 2016 *Journal of computer and systems sciences international* **55** 446
[9] Bjushgens G S (ed) 1998 Aerodynamics, Stability and Controllability of Supersonic Aircraft (Moscow: Nauka) p 816 (In Russian)

Isomorphic observers of the linear systems state

V S Kulabukhov

«Avionica» Joint Stock Company, Moscow, 127055, Russia

E-mail: kws0704@mail.ru

Abstract. The paper suggests isomorphic observers as a new type of observers. It is shown that isomorphic observers provide the maximum achievable quality of monitoring and filtering for linear systems. The techniques of synthesis of such observers are discussed. Isomorphic observers are multiplicative in form. This permits us to subdivide the general problem into separate tasks of the observer and filter synthesis. The means to achieve robustness and accuracy for isomorphic observers at high levels noises and parametric perturbations are also discussed. The example of the synthesis is presented.

1. Introduction

The most famous are two types of observers: the ones based on Kalman filter [1–4] and Luenberger observers [2, 3, 5]. They are applied mainly to linear systems and are asymptotic [1–5]. At the same time these observers carry out the function of signal filtering. However, the heuristic compromise has to be established between the qualities of observation and filtering. The lack of robustness of these observers is also eliminated heuristically [2, 3, 6].

Modern methods of observer synthesis are complex and poorly formalized. This applies to linear systems, which are often used in practice. Synthesis of observers for nonlinear systems is still more sophisticated [2, 3, 7–11].

The work [12] formally defines the structure of the new type of observers for general-form linear systems. These observers are called isomorphic [12, 13]. Based on the results of the works [12, 13], the techniques of synthesis of such observers are discussed. Isomorphic observers are multiplicative in form. This permits us to subdivide the general task into separate tasks of the observer and filter synthesis. It is demonstrated that isomorphic observers provide the linear systems with the maximum achievable quality of monitoring and filtering.

2. Linear isomorphic observers

The general-form linear system is considered

$$\begin{aligned} &\dot{X} = AX, X(0) = X_0, \\ &Z = HX, X \subset R^n, Z \subset R^m. \end{aligned} \tag{1}$$

In the Laplacian form (1) it is presented as follows

$$\begin{cases} X(p) = \Phi X(p), \\ Z(p) = HX(p). \end{cases} \tag{2}$$

Here $\Phi(p)=(pI-A)^{-1}$ is the fundamental matrix, I – the identity matrix, $Z(p)$ – the vector of direct measurements. It is required [1, 2] to obtain the estimate of the vector $X(0)=X_0$. The matrix $\Phi(p)$ is always inverse, i.e. is an isomorphism. If (2) is observable, then there is a commutative diagram [12, 13], shown in Fig. 1. If $m \geq n$, the observer S_N (2) is determined by the formula [12, 13]

$$S_N = \Phi^{-1}(p)H^{-1} = (pI-A)H^{-1} \tag{3}$$

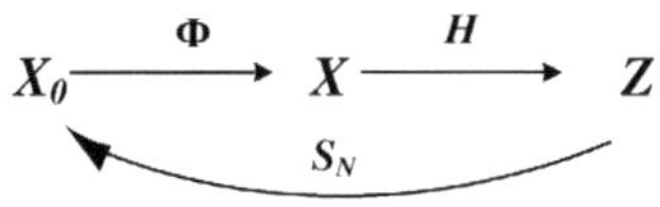

Figure 1. The projections composition chart for the observed linear system

Observers (3) are isomorphic [12, 13]. In the commutative diagram they establish a one-to-one correspondence between Z and X_0 with accuracy up to isomorphism Φ. It is shown [12, 13] that, for observed object, if $m \geq n$ there are always inverse $(S_N)^{-1}$ and H^{-1}. At the same time the matrices S_N and H can be rectangular and are non-invertible outside the chart. If the object is observable, and $m < n$, it is necessary to find another isomorphism, with the precision to which the observer can be calculated. In this case, it is always possible to transform (1) into the form with the invertible (isomorphic) matrix H^*. It connects the same outputs Z of the sensors with the vector of measured coordinates $X^* \subset X$, as in (1). The conversion is performed by expressing the unobserved coordinates through the measured. The possibility of such a transformation may serve as a criterion of observability.

Let the system be observable and $m \geq n$. We require that the value $\tilde{X}_0$ generated at the output of the observer satisfies the condition $\tilde{X}_0 \equiv X_0$. There are possible noises in the input signal X_0 and in the signal $\tilde{X}_0$.For their filtering we use the filter $W_\Phi = (pI-C)^{-1}D$, where C and D are matrices. The filter is selected by engineering quality criteria [14]. The equations (1) with the filter will take the form of

$$\begin{aligned} &X(p) = \Phi X_0, \\ &Z(p) = HX(p), \\ &\tilde{X}_0(p) = S_N Z(p), \\ &\hat{X}_0(p) = W_\Phi \tilde{X}_0(p). \end{aligned} \tag{4}$$

In the simplest case, if $\hat{X}_0 \equiv \tilde{X}_0$, $W_\Phi=I$, where I is the identity matrix. Fig. 2 shows the diagram for (4), which allows to transform (3) into the form

$$S_N=I\Phi^{-1}(p)H^{-1}=(pI-A)H^{-1},\ (S_N)^{-1}=H\Phi(p)I=H\Phi(p)=H(pI-A)^{-1}, \tag{5}$$

where $(S_N)^{-1}$ and H^{-1} satisfy the conditions of [12, 13]:

$$I_{X_0}^{left} = S_N(S_N)^{-1},\ I_{X_0}^{right} = \Phi^{-1}\Phi,\ I_{X_0}^{left} = I_{X_0}^{right} = I\ ,$$

$$I_Z^{left} = HH^{-1},\ I_Z^{right} = (S_N)^{-1}S_N,\ I_Z^{left} = I_Z^{right} = I_Z,\ I_Z I_Z = I_Z,\ I_Z = (I_Z)^{-1}. \tag{6}$$

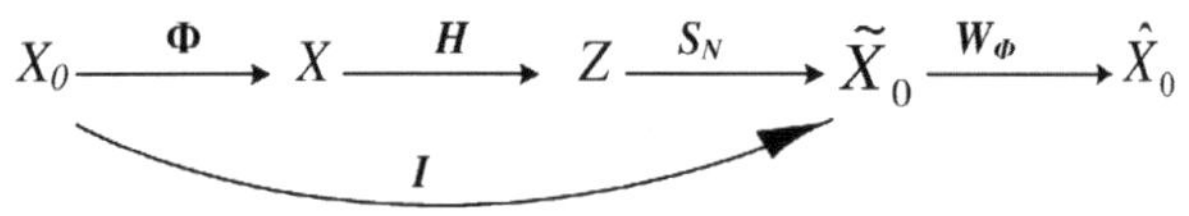

Figure 2. The projections composition chart for the system (4)

Here $I_{X_0}^{left}$, $I_{X_0}^{right}$ are the left and the right units on $\widetilde{X}_0 \equiv X_0$, I_Z^{left} and I_Z^{right} – the left and the right units on *Z*. Unit on $\widetilde{X}_0 \equiv X_0$ is equa lto the identity matrix *I*. In general [12, 13] I_Z^{left} and I_Z^{right} can be non-trivial matrices in the ordinary sense, but such that (6) holds true. The equations (5) and (6) allow us to determine the isomorphic observer S_N with the accuracy up to isomorphism Φ. If the model (2) corresponds exactly to the object (under the object we understand the object of observation, together with the sensors), the monitoring circuit with feedback (FB) S_N (see Fig. 1, 2) acquires a “linear” shape (see Fig.3). Taking (5) into account, it takes the form shown in Fig. 4.

$$X_0 \xrightarrow{\Phi} X \xrightarrow{H} Z \xrightarrow{S_N} \widetilde{X}_0 \xrightarrow{W_\Phi} \hat{X}_0$$

Fig. 3. "Linear" circuit with isomorphic observer

$$X_0 \xrightarrow{\textbf{Object}} Z \xrightarrow{S_N} \widetilde{X}_0 \xrightarrow{W_\Phi} \hat{X}_0$$

Fig. 4. "Linear" circuit of real object monitoring

Such a "linear" circuit meets the condition $S_N(S_N)^{-1}=I$ of precise tracking $\widetilde{X}_0 \equiv X_0$ at any point in time. S_N is the perfect observer: based on *Z* data it allows to observe the state X_0 instantly, fully, and accurately and hence also the state *X* at the time t under certain conditions $t_0 \to t$. The observer S_N is multiplicative, as it requires just matrix multiplication. As it can be seen from (5), Fig. 3 and Fig. 4, the result of the observation X_0 is completely determined by the choice of filter. The observer and the filter are synthesized separately. For the ideal observer the filter is formed under the given requirements to filtration quality. In this sense, the isomorphic observer is the ultimate in quality of signal X_0 monitoring and filtering. One should consider $W_\Phi=(pI-C)^{-1}D$ [13] as a type of filter, which, together with S_N forms physically realizable structures of monitoring and filtering.

In case of differences between the system and the model (2), (4), the result of the monitoring and filtering may be unsatisfactory. Then the "linear" circuit will need to be covered by additional feedback, as shown in Fig. 5.

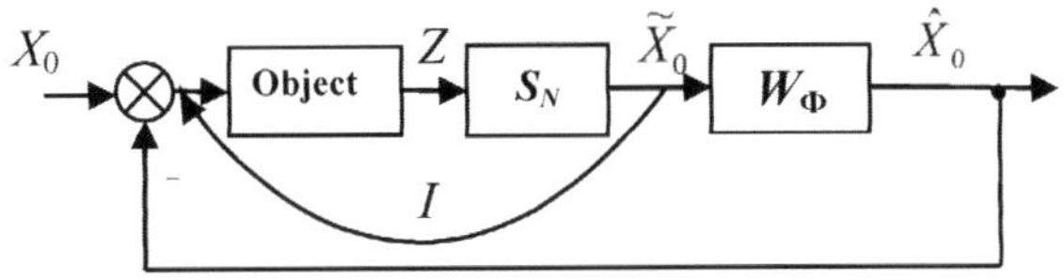

Figure 5. Robust monitoring circuit

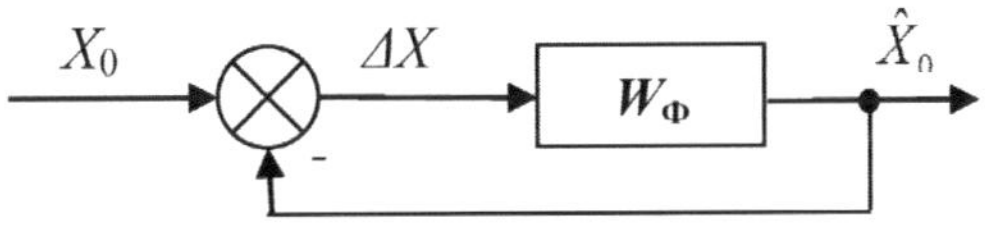

Figure 6. Robust circuit for the model, adequate to the object

For the model (4), adequate to the object, the transfer function "object-observer" from X_0 to $\widetilde{X}_0$ is close to $W_{OH} \approx I$. A closed circuit from X_0 to $\hat{X}_0$ (see Fig. 5) takes the form shown in Fig. 6. The transfer function of the closed circuit is $W=W_\Phi(I+W_\Phi)^{-1}$. That is, the characteristics of the robust circuit, as well as the "linear" one, are defined only by the structure of the filter. The astatism of the circuit is ensured, like in [15, 16], by the choice of W_Φ with high gain coefficient.

3. Example

Let $m<n$ and just one output be available for measurement in (2)

$$A=\begin{bmatrix} a_{11} & a_{12} \\ a_{21} & a_{22} \end{bmatrix}=\begin{bmatrix} -4 & -1{,}14 \\ 1 & -0{,}62 \end{bmatrix}, \mathrm{H}=\begin{bmatrix} h_{11} & 0 \end{bmatrix}=\begin{bmatrix} 1 & 0 \end{bmatrix}. \tag{7}$$

Write (2) with regard to (7) in scalar form

$$\begin{cases} x_1=(p-a_{11})^{-1}a_{12}x_2+(p-a_{11})^{-1}x_{10}, \\ x_2=(p-a_{22})^{-1}a_{21}x_1+(p-a_{22})^{-1}x_{20}, \\ z=h_{11}x_1. \end{cases} \tag{8}$$

By substitution of the second equation to the first one and some other evident substitutions, we will get the estimation $\tilde{x}_{20}$:

$$\tilde{x}_{20}=\left[(p-a_{11})\, a_{12}^{-1}(p-a_{22})-a_{21}\right]h_{11}^{-1}z-a_{12}^{-1}(p-a_{22})x_{10}=a_{20}^{-1}h_{11}^{-1}z+a_{10}x_{10}, \tag{9}$$

where $a_{20}^{-1}=[(p-a_{11})\, a_{12}^{-1}(p-a_{22})-a_{21}], a_{10}=-a_{12}^{-1}(p-a_{22})$. The observer in (9) is calculated with accuracy up to isomorphism h_{11} using the formula

$$S_N=a_{20}^{-1}h_{11}^{-1}. \tag{10}$$

Thus,

$$\tilde{x}_{20}=a_{10}x_{10}+S_N z \tag{11}$$

In the observation equation (11), as well as in the Kalman filter, the updating sequence characterized by the second addend $S_N z$, is formed by mesurements. Modeling of the "linear circuit" (see Fig. 4) with the observer (10), (11) and a filter $W_\Phi=k_\Phi(T_\Phi p+1)^{-1}$ with $k_\Phi=1$ and $T_\Phi=0{,}4$ has shown that the estimation $\hat{x}_2$ tracks the change x_2 accurately in accordance with the dynamics of the filter. With inaccurate knowledge of the elements a_{ij} in (7), (8) an error appears in the "linear" circuit in observation x_2. Closing the circuit with an additional feedback in accordance with Fig. 5–6 provides the robustness. Linear and robust circuits were also simulated with filters [13]

$$W_\Phi=\frac{1}{0.4p+1}\cdot\frac{1}{1.5p+1}$$ – for "linear" circuit, and

$$W_\Phi=\frac{1}{0.4p+1}\cdot\frac{99}{1.5*100p+1}$$ – for a robust closed circuit.

Fig. 7 illustrates graphs of estimation and filtering of noisy input x_2 (figure 1), with the presence of normal white noise. The amplitude of the noise was 10% of the amplitude of the useful signal.

Graph 2 corresponds with the estimation $\hat{x}_2$ at the output of the "linear" circuit at the level of distortion of 40% of the parameter a_{11} in the object (7), (8). Graph 3 illustrates the estimation $\hat{x}_2$ at the output of robust circuit when the distortion of the parameter a_{11} is equal to 40%.

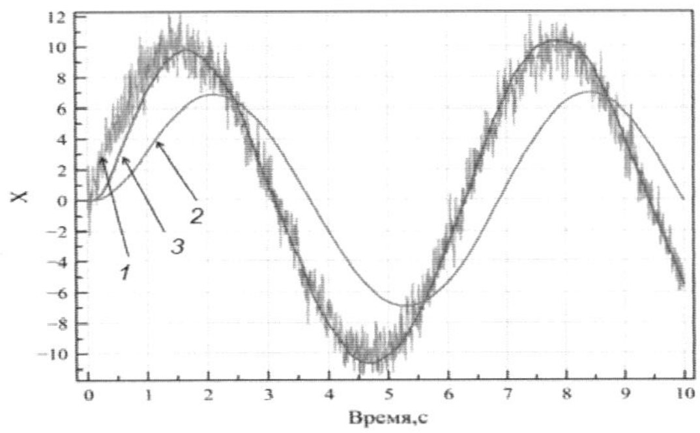

Figure 7. Estimation and filtering of noisy signal

It is seen that the "linear" circuit is subject to the influence of parametric perturbations, however it possesses a high-quality signal filtering. The circuit with an additional feedback possesses the property of robustness to significant variations (not known in advance) of the object parameters and a high-quality filtering of noises present in the input signal. If the deviations from the nominal values are known in advance (like systematic errors of the sensors), the "linear" circuit can be calibrated by selection of the filter parameters and elimination of their influence. The examples of estimating systematic errors in case of an aircraft are given in [17,18]. Additional feedback is needed in the case of significant deviations of the object parameters from those adopted in the model. The astaticism of the circuit can be adjusted by changing the gain coefficient of the filter.

So we conclude that in this paper the synthesis of isomorphic regulators is formalized to the maximum. High quality monitoring and filtering is ensured by filter selection taking into account the physical realizability of the system "observer-filter" and the desired properties of signal filtering. In that sense, isomorphic observers are "marginal" in terms of the achievable quality of estimation and filtering in comparison with well-known observers.

References

[1] Kalman, R, Falb P and Arbib M 1971 *Essays on mathematical systems theory* (Moscow: Mir)
[2] Krasovsky A A 1987 *A handbook on the theory of automatic control* (Moscow: Nauka)
[3] Krasnova S A, Utkin V A 2006 *Cascade synthesis of observers of dynamic systems state* (Moscow: Nauka)
[4] Shahtarin B I 2008 *Wiener and Kalman filters.* (Moscow, Gelios ARV)
[5] Luenberger D B 1966 *IEEE Trans Automat. Control* **11** 190
[6] Chernodarov A V 2017 *Control, diagnostics and identification of aircraft devices and measuring and computing complexes* (Moscow: "Nauchtechlitizdat" LLC).
[7] Isidori A 1995 *Nonlinear control systems* (Berlin: Springer-Verlag).
[8] Ciccarella G, Dalla Mora M and German A A 1993 *Int. J. Control* **57** 537
[9] Proychev Ph and Mishkov R L 1993 *Automatica* **29** 495
[10] Shouse K R, Taylor D G 1993 *IEEE Trans. Automat. Control* **29** 95
[11] Wanem W M 1980 *Linear multivariable control systems. Geometrical approach* (Moscow)

[12] Kulabukhov V S 2014 The principle of isomorphism in the implementation task and its application to the analysis of properties of control systems *Proc. XII all–Russian meeting on control problems* (Moscow: ICS RAS) 438–448.
[13] Kulabukhov V S 2017 *Radiotekhnika* **8** 50
[14] http://portal.tpu.ru/SHARED/v/VOS/study/disc1/Tab/tema06.pdf (date of access - 16.05.2017)
[15] Kulabukhov V S 2017 *Mechatronics, automation, control* **18** 507
[16] Kulabukhov V 2017 *MATEC Web of Conferences* **99** 03008.
[17] Korsun O N and Poplavskii B K 2011 *Journal of computer and systems sciences international* **50** 130
[18] Korsun O N, Nikolaev S V and Pushkov S G 2016 *Journal of computer and systems sciences international* **55** 446

Extreme trajectory approach to the problem of determining the degree of stability of a laser surface scanning system

V N Nikitin and V B Chemodanov

Flight Dynamics and Control Department, Moscow Aviation Institute (National Research University), Moscow, 125993, Russia

E-mail: marmottoduo@yandex.ru

Abstract. The degree of stability of a laser system for surface scanning with nonlinear multiplicative crosstalks is discussed. To determine its stability, the action functional is introduced, which is defined on the set of virtual (achievable) trajectories. The action functional is a measure of external action, which should be applied to a system to move it along a predetermined trial trajectory in the state space.The degree of stability of the system depends on the minimum value of the action functional which is reached on the extreme trajectory transferring the laser scanning system from equilibrium to the limit of the normal operation range. Numerical methods are proposed for calculating the degree of stability.

1. Introduction

Laser surface scanning systems (LSS) are used to solve many scientific and operational problems [1-4]. The most famous producers of laser surface scanners are Cambrige Technology in the USA and SCANLAB AG in Germany. Fig. 1 shows the functional diagram of a laser scanner from the German company. The cantilever arrangement of mirrors in most of the scanners produced today causes the cantilevers to wobble. Nonlinear crosstalks [5-6] arise as a result between beam steering channels of multiplicative type: the greater is the deflection of a mirror in one channel, the greater is the influence on the other channel. Similar problem results from building an LSS using one electromechanical two-axis deflector instead of two, whose layout is shown in Fig. 2. [7].

Our theoretical studies and experiments have shown that the above nonlinear effects narrow the area near the equilibrium position $x_1 = x_2 = 0$. In particular, self-oscillations can arise for certain deviation values $h = \sqrt{x_1{}^2 + x_2{}^2}$, which may be detected or show themselves in the space of phase coordinates by the emergence of the limit cycle.

2. Action functional

The action functional $S[\boldsymbol{y}(t)]$ is defined as a time integral from of the square of the difference of time derivative with respect to $\boldsymbol{y}(t)$ and the right-hand part of the dynamical equation of the system [8-9]. Let us make this definition more specific taking a nonlinear autonomous system of the n-th order with the control action vector $\boldsymbol{u}(\boldsymbol{x})$ as an example:

$$\frac{dx}{dt} = \boldsymbol{b}(\boldsymbol{x}, \boldsymbol{u}(\boldsymbol{x})) \equiv \boldsymbol{a}(\boldsymbol{x}); \quad \boldsymbol{x} = (x_1, x_2, \dots, x_n)^T. \quad (1)$$

Suppose the point $\boldsymbol{x} = 0$ is the equilibrium, that is $\boldsymbol{a}(0) = 0$. It is assumed for $\boldsymbol{a}(\boldsymbol{x}) = (a_1, a_2, \dots, a_n)^T$ that in the domain $O = \{\boldsymbol{x} \mid \| \boldsymbol{x} \| \leq h\}$, where $h > 0$ is a constant value, the vector function $\boldsymbol{a}(\boldsymbol{x})$ is piecewise continuous and it allows a unique solution to the system (1) for each initial state $\boldsymbol{x}_a \in \boldsymbol{O}$.

Let us define the following functional on the set $C_{0T}(\boldsymbol{x}_{in}, \boldsymbol{x}_f)$ of virtual trajectories $\boldsymbol{y}(t)$ belonging to the space of functions given and continuous on the segment t_{in}, t_f with the range of values in R^n and such that $\boldsymbol{y}(t_{in}) = \boldsymbol{x}_{in}$, $\boldsymbol{y}(\boldsymbol{t}_f) = \boldsymbol{x}_f$:

$$S[\boldsymbol{y}(t)] = \frac{1}{2}\int_0^T \left[\dot{\boldsymbol{y}}(t) - \boldsymbol{a}(\boldsymbol{y}(t))\right]^2 dt, \quad (2)$$

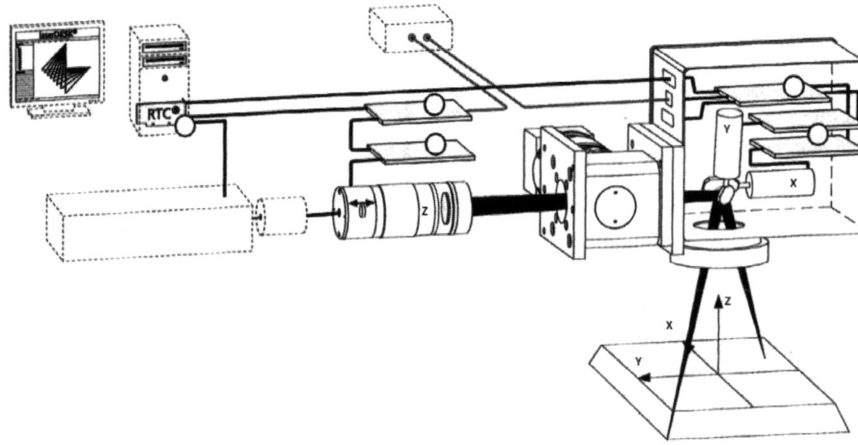

Figure 1. The functional diagram of a laser scan unit.

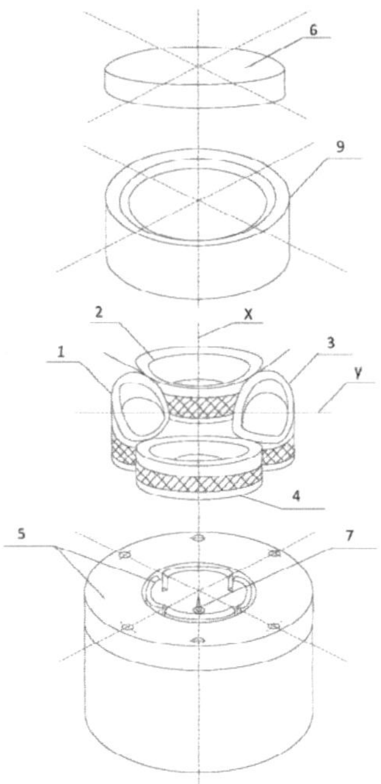

Figure 2. The layout of an electromechanical two-axis deflector:

1, 2, 3, and 4 - coils, 5 – magnetic system, 6 – deflecting mirror, 7 – support, 8 – mount.

Consider the physical meaning of the functional $S[\boldsymbol{y}(t)]$. If the virtual trajectory $\boldsymbol{y(t)}$ does not match the solution $\boldsymbol{x}$ *(t)* to the equation (1), then the movement along $\boldsymbol{y(t)}$ is described by the equation

$$\frac{dy(t)}{dt} = \boldsymbol{a}(\boldsymbol{y}(t)) + \boldsymbol{r}(t) \quad \text{or} \quad \boldsymbol{r}(t) = \dot{\boldsymbol{y}}(t) - \boldsymbol{a}(\boldsymbol{y}(t)), \tag{3}$$

where $\boldsymbol{r(t)}$ is the external action that should be applied to the system described by the equation (1) for transition from the initial state $(\boldsymbol{x_{in}}, \boldsymbol{t_{in}})$ to the final state $(\boldsymbol{x_f}, \boldsymbol{t_f})$ during the time $T = t_f - t_{in}$ along the trajectory $\boldsymbol{y}(t)$.

Thus the action functional $S[\boldsymbol{y}(t)]$ is the square of the norm given in the space of functions integrable with the square of $L_2[t_{in}, t_f]$ for the external action $\boldsymbol{r(t)}$ that moves the system from the initial state $(\boldsymbol{x_{in}}, \boldsymbol{t_{in}})$ to the final state $(\boldsymbol{x_f}, \boldsymbol{t_f})$:

$$S[\boldsymbol{y}(t)] = \frac{1}{2}\|\boldsymbol{r}(t)\|^2. \tag{4}$$

3. The degree of stability

Let us define the degree of stability $U_F(\boldsymbol{x}_0)$ for a system that is in equilibrium at the point $\boldsymbol{x} = \boldsymbol{x_0} = \boldsymbol{0}$ lying within a domain $\mathcal{D}$ where $\mathcal{D}$ is the normal operation range as the minimum value of the square of the norm (see (5)) in the space of $L_2[t_{in}, t_f]$ for the external action that takes the system out of equilibrium to the bound $\boldsymbol{F}$ of the normal operation range of the system.

If the trajectory $\boldsymbol{x}(t)$ is within $\mathcal{D}$, that is $\boldsymbol{x}(t) \subset \mathcal{D}$, then the system functions normally. When $\boldsymbol{x}(t)$ goes beyond $\mathcal{D}$, an emergency situation arises, which is called "control failure" in technical systems. By definition

$$U_F(\boldsymbol{x}_0) = \min_{\boldsymbol{x}_f \in F} \inf_{\boldsymbol{y}(t) \in C_{0T}(\boldsymbol{x}_{in}, \boldsymbol{x}_f)} \{S[\boldsymbol{y}(t)] | \boldsymbol{y}(0) = \boldsymbol{x}_{in}; \boldsymbol{y}(T) = \boldsymbol{x}_f; T - \text{var}\} (\boldsymbol{x}_0, \boldsymbol{x}_f)$$

and so the degree of stability of $U_F(\boldsymbol{x}_0)$ for $\mathcal{D}$, depends on the minimum value of the quasipotential $V(\boldsymbol{x}_0, \boldsymbol{x}_f)$ among all points $\boldsymbol{x}_f$ belonging to the bound F of $\mathcal{D}$.

4. The numerical method of defining the degree of stability

It should be noted that the search for the exact greatest lower bound of the functional $S[\boldsymbol{y}(t)]$ in (4) is a difficult problem. It can be solved approximately, for example, by approximating trial trajectories $\boldsymbol{y}(t)$ with polygonal lines $\boldsymbol{y}^{(m)}(t)$ having $m - 1$ salient points $(\boldsymbol{y}_1, t_1)$, $(\boldsymbol{y}_2, t_2)$... $(\boldsymbol{y}_m, t_m)$ where $t_{j+1} = t_j + \Delta t$, $\Delta t = (t_f - t_{in})/m$,

The equation of the polygonal line $\boldsymbol{y}_w^{(m)}(t)$ will be written as follows:

$$\boldsymbol{y}_w^{(m)}(\theta) = \boldsymbol{y}_j + (\boldsymbol{y}_{j+1} - \boldsymbol{y}_j)(\theta - j), \qquad \text{for } j \le \theta \le j+1$$

where $\theta = \frac{mt}{T}$ and j is an integer parameter that assumes values 0, 1, ..., m-1. In addition to $\boldsymbol{y}_w^m(t)$ in (4), other types of trajectories $\boldsymbol{y}_s^{(m)}(t)$ based on a given complete system of functions can be used, for example, a set of harmonic functions with *m* harmonics. Furthermore, it is very important to have a set of exact analytical expressions to be able to debug software and for the illustration of convergence of the greatest lower bound of the functional $S[\boldsymbol{y}_m(t)]$ to the greatest lower bound of $S[\boldsymbol{y}(t)]$ with m $\to \infty$.

5. Degree of stability of a laser surface scanning system

To illustrate the numerical method proposed in section 4 above, consider the laser surface scanning system with nonlinear crosstalks of multiplicative type, whose block diagram is shown in Fig. 3.

The system of dynamical equations corresponding to this figure has the following form:

$$\dot{x_1} = K_d x_2; \; \dot{x_2} = \frac{C_M}{J} x_3;$$

$$\dot{x_3} = \frac{1}{L}[-Rx_3 - (KK_T + C_E)x_2 - K_y K x_1 + \frac{K_{\text{пс}}}{K_d} x_5 x_1 x_4];$$

$$\dot{x_4} = K_d x_5; \; \dot{x_5} = \frac{C_M}{J} x_6; \tag{5}$$

IOP Conf. Series: Materials Science and Engineering **312** (2018) 012017 doi:10.1088/1757-899X/312/1/012017

$$x_6 = \frac{1}{L}[-Rx_6 - (KK_T + C_E)x_5 - K_yKx_4 + \frac{K_{пс}}{K_d}x_2x_1x_4].$$

The action functional (3) for (7) if m=1 has the form:

$$S[\vec{y}_w^1(t)] = \frac{1}{2}\int_0^T \left\{ (\frac{x_1}{T} - K_d\frac{t}{T}x_2)^2 + (\frac{x_2}{T} - \frac{C_M}{J}\frac{t}{T}x_3)^2 + \left(\frac{x_3}{T} + \left[\frac{R}{L}x_3 + \frac{KK_T+C_E}{L}x_2 + \frac{KK_y}{L}x_1\right]\frac{t}{T} - \frac{K_{пс}}{K_dL}x_4^2x_1\frac{t^2}{T^3}\right)^2 + \left(\frac{x_4}{T} - K_d\frac{t}{T}x_5\right)^2 + \left(\frac{x_5}{T} - \frac{C_M}{J}\frac{t}{T}x_6\right)^2 + \left(\frac{x_6}{T} + \left[\frac{R}{L}x_6 + \frac{KK_T+C_E}{L}x_5 + \frac{KK_y}{L}x_4\right]\frac{t}{T} - \frac{K_{пс}}{K_dL}x_1^2x_4\frac{t^2}{T^3}\right)^2 \right\} dt. \quad (6)$$

Examples of the D domain forms are shown in Fig. 4. The case shown in the left-hand part of Fig. 4 corresponds to the problem of a system going to the stops located at $x_1 = \pm\Delta$ and $x_4 = \pm\Delta$.

The domain D_2 is characterized by σ that defines an acceptable deviation of the system from the origin of coordinates. The degree of stability of the system is characterized by U equal to the minimum value of the $S[\boldsymbol{y}(t)]|\boldsymbol{y}(0) = \boldsymbol{0}; \boldsymbol{y}(T) = \boldsymbol{x}_f$ wthere $\boldsymbol{x}_f$ lay on the bound F of the domain D.

If the crosstalk coefficient $K_{пс} = 0$, then the right-hand part of expression (5) is a quadratic form in x_i, $i = \overline{1,6}$. The minimum in the right-hand part of (6) can be achieved for $x_4^* = x_5^* = x_6^* = 0$ (the second channel is not disturbed and it does not affect the degree of stability of a single-channel device). The values of x_2^* and x_3^* are expressed analytically by T. The case of $K_{пс} = 0$ can be used, therefore, as a test example. Fig. 5.6 shows the calculated dependence of the degree of stability of a system $U^{(1)}(0; K_c) / U^{(1)}(0; 0)$ on $K_c = K_{пс}$ obtained in the first-order $m = 1$ approximation where $U^{(1)}(0; 0)$ is the degree of stability of the system for $K_c = 0$. The calculations were made for the following parameter values:

$$K_{d=57}, L = 10^{-2}, C_M = 6.2, J = 4.5 \times 10^{-2};$$

$$C_E = 6.2, K_T = 0{,}0217;\ K_y = 5.13, k = 10, \Delta = 3$$

The data thus obtained imply that the stability of the system decreases as $K_{пс}$ increases. The $m = 1$ approximation naturally allows to produce only qualitative results. Therefore, it is interesting to consider higher orders of approximation to get more accurate results.

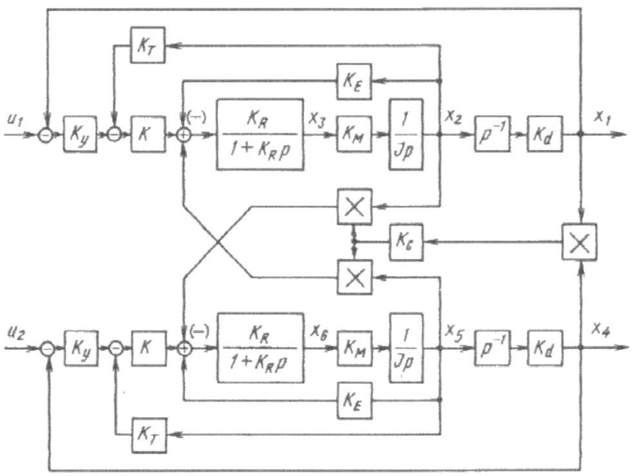

Figure 3. The block diagram of a laser surface scanning system.

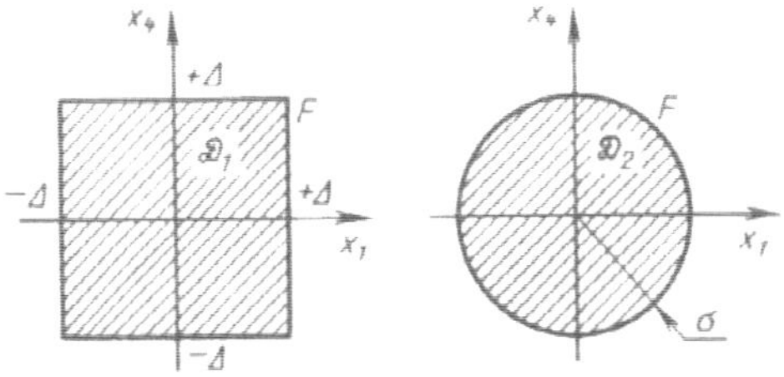

Figure 4. Examples of the form of the normal operation range of a laser surface scanning system.

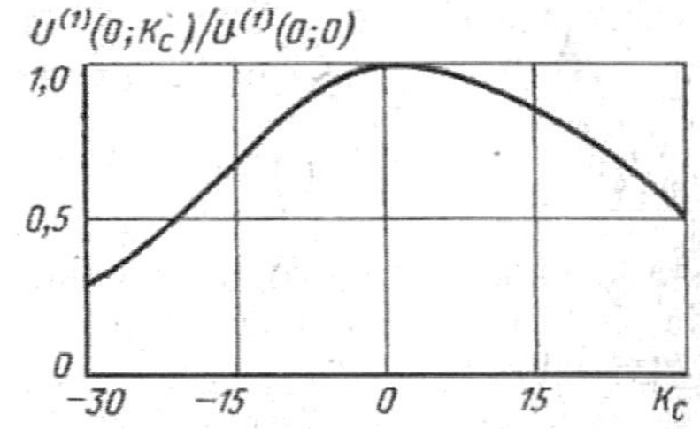

Figure 5. The degree of stability of a laser system for surface scanning in the first-order approximation of m = 1 versus crosstalk coefficient $K_C = K_{\Pi C}$.

References

[1] Marshall F G and Stutz G E 2004 *Handbook of Optical and Laser Scanning*, (New York: Marcel Dekker)
[2] Abuazzaa A, Brabazonb D and El-Baradieb M A 2005 *Multi-Beam Fiber Optic Lazer Scanning System for Surface Defect Recognition* (Tripoli: Physics Department Al-Fatha University)
[3] Quan C and Wang S H 2000 *Optical Communication* **183** 19-27.
[4] His-Yung F, Yixin L and Fengfeng X 2001 *Precision Engineering* **25** 185-191.
[5] Llorente R, Clavero R, Ramos F, and Marti J 2005 *EURASIP Journal on Applied Signal Processing* **10** 1593–1602.
[6] Seiji N and Kenichi I 2006 *Electronics and Communications in Japan* **89** 50–64
[7] Chemodanov V 2016 *Automatic Systems and Laser Beam Guiding Devices* (Moscow: FIZMATLIT)
[8] Woan G 2010 *The Cambridge Handbook of Physics Formulas* (Cambridge: University Press)
[9] Hand L N and Finch J D, 2008 *Analytical Mechanics* (Cambridge: University Press)
[10] Aström K J, Albertos P, Blanke M, Isidori A, Schaufelberger W and Sanz R 2001 *Control and Complex Systems* (London: Springer-Verlag)
[11] Blais F 1987 *Optomechanical Systems Engineering* SPIE **0817** 14

IOP Conf. Series: Materials Science and Engineering **312** (2018) 012018 doi:10.1088/1757-899X/312/1/012018

Construction of multi-agent mobile robots control system in the problem of persecution with using a modified reinforcement learning method based on neural networks

M L Patkin and G N Rogachev

Samara State Technical University, Samara, Russia

E-mailpatkinm@gmail.com and grogachev@mail.ru

Abstract. A method for constructing a multi-agent control system for mobile robots based on training with reinforcement using deep neural networks is considered. Synthesis of the management system is proposed to be carried out with reinforcement training and the modified Actor-Critic method, in which the Actor module is divided into Action Actor and Communication Actor in order to simultaneously manage mobile robots and communicate with partners. Communication is carried out by sending partners at each step a vector of real numbers that are added to the observation vector and affect the behaviour. Functions of Actors and Critic are approximated by deep neural networks. The Critics value function is trained by using the TD-error method and the Actor's function by using DDPG. The Communication Actor's neural network is trained through gradients received from partner agents. An environment in which a cooperative multi-agent interaction is present was developed, computer simulation of the application of this method in the control problem of two robots pursuing two goals was carried out.

1. Introduction

Complicating technical systems and tightening the requirements for the quality of their management lead to the need for continuous improvement of control algorithms. Methods of intellectual controlling and making decisions have been significantly improved recently due to the use of artificial neural networks.

Training with reinforcement is a very powerful method of solving problems of managing complex systems, the main advantage of which is that it does not require a prior knowledge of the control object. This knowledge arises in the process of interaction of the managing agent with the control object. In tasks that are solved in a multi-agent way the complexity of management is determined not only by the complexity of the management objects, but also by the level of agent interaction, which opens up almost limitless scope for the improvement of managing agents.

In this paper, we propose an environment for simulating multi-agent cooperative interactions and a method for educating agents to develop and understand messages about how to achieve the goal.

2. The task of optimal control of a group of robots

Let us consider the problem of optimal control [1] of a group of robots in the setting of cooperative partially observable Markov game [2]. Model of control objects is set:

$$x^i = f^i(s^i, a^i), \tag{1}$$

where $x^i \in R^{n_i}$; $s^i = [s_1^i ... s_{m_i}^i]$ - state vector of robot i; $a_i \in A_i \subseteq R^{m_i}$, $a^i = [a_1^i .. a_{n_i}^i]$ - vector of available actions for each robots; A_i - compact set; $i = 1,...,N$; N - number of robots.

Initial conditions:

$$x^i(0) = x^{0,i}, i = 1,...,N. \tag{2}$$

Group of robots aspire to achieve the set of goals. The number of goals equals to the number of robots in a group. Goals is the objects which are set by:

$$g^i = f_g^i(P_{norm}), \tag{3}$$

where P_{norm} represents normal probability distribution.

Terminal conditions:

$$\phi_k^i(x^i(t_g), g^i(t_g)) = 0, k = 1,...,l_i, l_i \le n_i, \tag{4}$$

where

$$t_g = \begin{cases} t, \; if \; t < t^{term} \; and \; \phi_k^i(x^i(t), g^i(t)) = 0, k = 1,...,l_i, i = 1,...,N \\ t^{term}, \; if \; t = t^{term}; \end{cases} \tag{5}$$

t^{term} - robots control time limit.

Constraint on the state:

$$\alpha_k(x^i(t), g^i(t)) \le 0, k = 1,...,r. \tag{6}$$

From (4) it is clear that for the robot i on the terminal condition will only depend on the state of robot i and the state of the goal i and certainly on global conditions such as (5) and (6). Simply speaking robot i is to catch the goal i.

For each robot it is necessary to find such a set of actions $[a_0,...,a_{t_g}]$ that $\phi_k^i(x^i(t_g), g^i(t_g)) = 0$.

3. Multi-agent Actor-Critic method for cooperative environment

The classic Actor-Critic method assumes that the policy function of the Actor will be formed based on value function from the Critic [3, 4]. The function of the Critic is developed according to the TD algorithm [5, 6]. In multi-agents cases (figure 1), especially in the case of POMDP, cooperation between agents begins to have a strong impact on the quality of the agents' work, because in observation i information that is more important for robot j can be found. For such situations the modification of the Actor-Critic algorithm was made.

The main idea of the modified Actor-Critic method for the case of interacting multi-agent systems in the division of the Actor into the Action Actor and the Communication Actor (figure 2). The Action Actor, as in the classic method, is responsible for the development of control actions, and the Communication Actor generates a message for the partners, which leads to more efficient management of the multi-agent system.

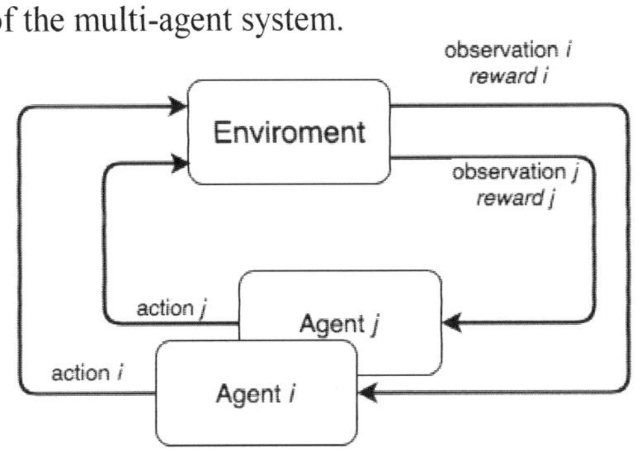

Figure 1. Multi-agent reinforcement learning.

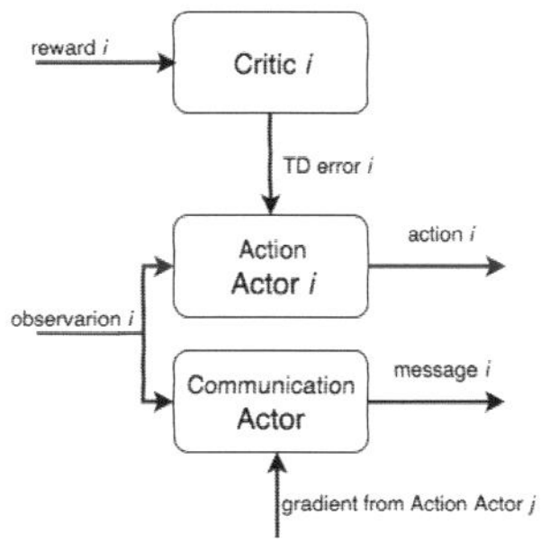

Figure 2. Scheme of modified Actor-Critic

Let us consider how the modified method works. On the step $t-1$, Agent i generates a message m_i that is added to the observation vectors o of the other agents (figure 3). On the basis of this, the remaining agents on the step t generate the control action a. In the process of learning the neural networks of the Actors, the propagation of gradients [7] that comes from the end of the neural network to the beginning, comes from the neural network of the Action Actor to the neural network of the Communication Actor. Under the action of gradients, the Communication Actor trains to generate such messages that would lead to a reduction in the Action Actor errors.

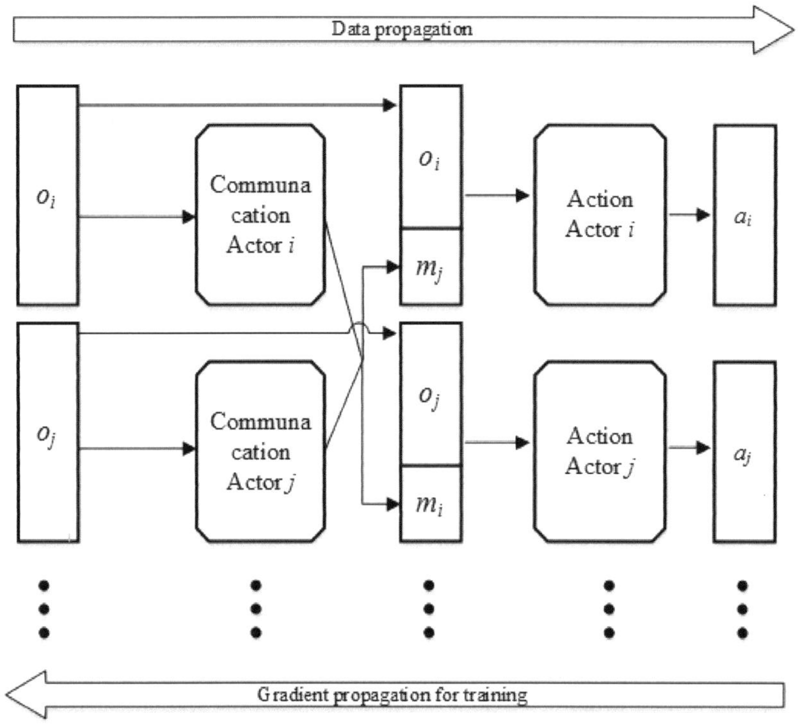

Figure 3. Modified Actor-Critic method.

4. Experiments

The environment is a field of 1000x1000 pixels, in which there are two robots and two targets (the environment code is based on https://github.com/harvitronix/reinforcement-learning-car). The objectives of robots: for the robot 1 it is to catch the target 1, for the robot 2 it is to understand the robot 2. The robots are equipped with device that indicate self position and position of the target, as well as with 3 sonars placed under different angles (figure 4). The goal of the robots: robot 1 is to catch target 1, robot 2 - goal 2. The target is considered caught if the robot approached it closer than 60 pixels (between the centers).

The robot is also able to "hear" what the partner says to it. The state of the robot is a vector of 9 numbers: 3 numbers for the sonars readings, 4 numbers for the position of the robot and the goal, and 2 numbers for the messages from the partner.

The robot is controlled by selecting one of three commands: "left", "forward", "right". The main idea of the experiment is that robot 1 observes goal 2, while it sis supposed to catch target 1, and vice versa. This feature allows you to find out how well agents can start communicating. Robots should learn by means of a communication channel to convey information about where the target is to the partner. Experiments were conducted for static purposes and for purposes that move with variable speed and in a random direction.

The figure 4 shows green and red circles - agents, white lines emanating from them - sonars, blue and yellow circles - targets. Table 1 shows architectures of neural networks of Actors and Critic where N_a - number of actions and N_m length of message vector.

Table 1. Architectures of Actors and Critic.

	Action Actor	Communication Actor	Critic
1	Fullyconnected: 256	Fullyconnected: 256	Fullyconnected: 256
2	Dropout: 20%	Dropout: 20%	Dropout: 20%
3	RELU	RELU	RELU
4	Fullyconnected: 256	Fullyconnected: 256	Fullyconnected: 256
5	Dropout: 20%	Dropout: 20%	Dropout: 20%
6	RELU	RELU	RELU
7	Fullyconnected: N_a	Fullyconnected: N_m	Fullyconnected: 1

Table 2. Experiments results. Sum of agents reward for 500 runnings.

	Static goals	Dynamic goals
No communication	-3314	-16777
Communication	37624	26797

Figures 5-8 show the results of trained agents. As it can be seen, the capture of targets was made from different positions and not in an optimal way. This non-optimality can be explained by the fact that the messages transmitted by partners did not always reflect how to act.

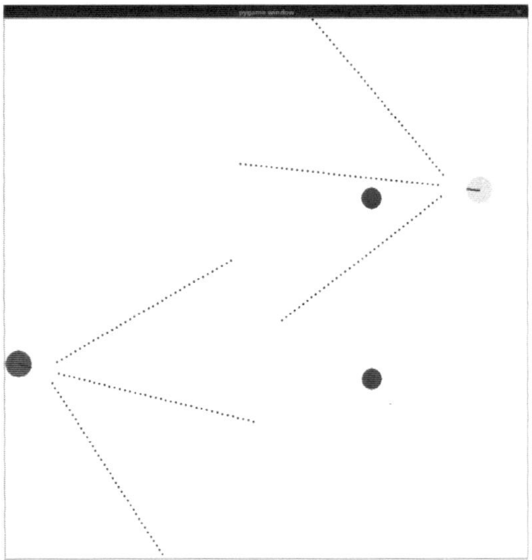

Figure 4. Visualisation of environment.

However, as it is shown in table 2, the result of the system trained by using the communication channel exceeds the result for the system without the communication. In this table the cells contain the values of the total reinforcement for 500 starts for two agents.

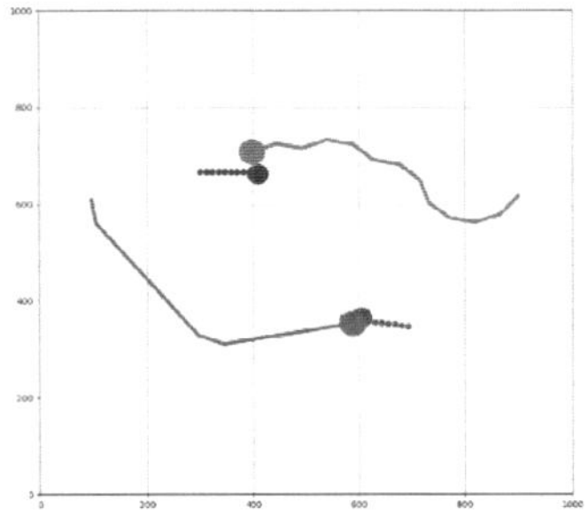

Figure 5. Experiment 1.

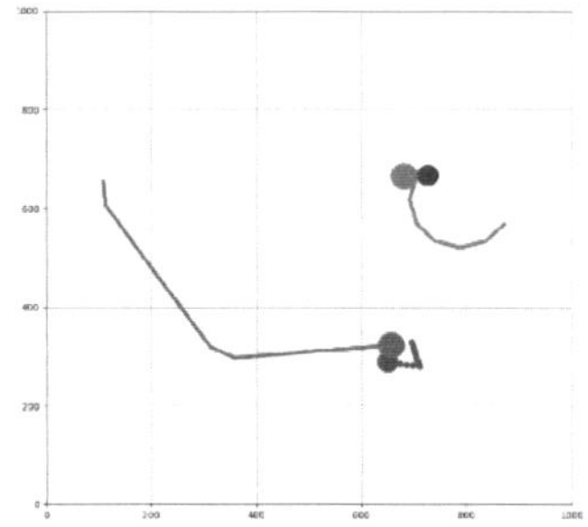

Figure 6. Experiment 2.

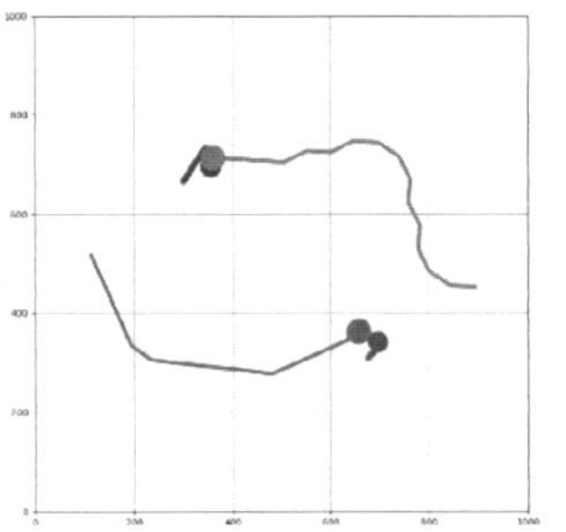

Figure 7. Experiment 3.

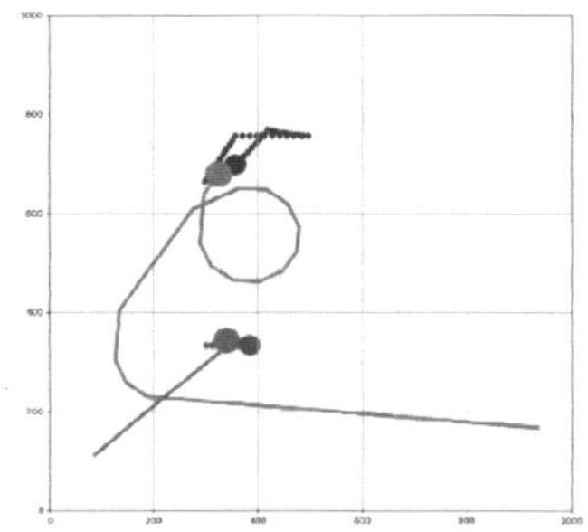

Figure 8. Experiment 4.

5. Conclusion

A modified neural network of the Actor-Critic method for controlling a multi-agent system in the form of interacting mobile robots is proposed. Similar work is currently underway [8, 9], which shows the promise of such an approach. In the future work on improving the quality of the learning algorithm is planned.

Acknowledgments

The reported study was funded by RFBR and Samara Region according to the research project № 17-48-630410.

References

[1] Rogachev G, Patkin M, and Rogachev N 2017 *ITM Web of Conferences* **10** 02006.
[2] Witten I H 1977 *IInformation and control* **34**(4) 286.
[3] Konda V R and Tsitsiklis J N. 2000 Actor-critic algorithms *Advances in neural information processing systems.*
[4] Bhatnagar S et al. 2009 *Automatica* 45(11) 2471.
[5] Tesauro G 1995 *Communications of the ACM* **38** 58.
[6] Sutton R S, and Barto A G 1998 *Reinforcement learning: An introduction* (Cambridge: MIT press).
[7] Silver D et al. 2014 "Deterministic policy gradient algorithms." *Proceedings of the 31st International Conference on Machine Learning (ICML-14).*
[8] Gupta J K, Egorov M and Kochenderfe M 2017 "Cooperative multi-agent control using deep reinforcement learning." *International Conference on Autonomous Agents and Multiagent Systems.* Springer, Cham.
[9] Lowe R et al. 2017 "Multi-Agent Actor-Critic for Mixed Cooperative-Competitive Environments." *arXiv preprint arXiv*:1706.02275

Optimization of maintenance period for the elements of resource-supplying networks with branching structure

A I Peschansky[1], G N Rogachev[2 *] and A I Kovalenko[2]

[1]Sevastopol State University, Sevastopol, Russia
[2] Samara State Technical University, Samara, Russia

E-mail: grogachev@mail.ru

Abstract. A restorable system with linear branching structure and finite reliability is considered. Preventive maintenance of its elements is carried out. Operation and restoration periods are assumed to be random values of general kind. Iteration processes of calculation of stationary reliability and economical characteristics of the network are constructed. Optimal intervals between elements' maintenance are obtained as the functions of their times-to-failure. The examples of electricity supply network and commutation network equipment at the enterprise are given. The optimal choice of elements' preventive maintenance is shown to result in economical network indexes improvement.

1. Introduction

Practical necessity has resulted in a considerable widening of studies in the theory of complex technical systems' reliability, specifically resource-supplying networks. The problems of reliability are quite many-sided. They are connected with technological, constructive, organizational aspects. The necessity of development of fundamental mathematics adapted for specific problems in reliability is evident. But in most cases calculation formulas are obtained under assumption of exponential distribution laws of restoration and operation times of elements. In this case Markov random processes are applicable.

Considering general kinds of the above variables complicates the problem of getting reliability and economical characteristics of systems considerably. And recurrent algorithms of calculating stationary characteristics of resource-supplying networks with branching structure are constructed in this type of problem setting. Optimal period of preventive maintenance is obtained.

2. Problem setting

Let us consider resource-supplying network having a many-layered linear branching structure in which every element of some layer is over several elements of lower layer. Each element is connected with the elements of next layers only [1]. The structure of such a network is given in Fig. 1.

The main element a_0 is connected with a_1 elements of the first layer, each of which, in its turn, is connected with a_2 elements of the second level and so on. Each element of the second to last $(n-1)^{\text{th}}$ layer is connected with a_n layers of the last n^{th} layer. The elements of the last layer are called output ones.

Let us describe the operation of the above network. The failure of any element of the network is discovered immediately and at once its restoration begins. At the moment of failure, both operation and restoration of an element are stopped. The network is supposed to be homogeneous, i.e. the elements of one layer are the same. Time-to-failure of elements of ith layer is a random variable α_i with the distribution function $F_i(t) = P(\alpha_i \le t), i = \overline{0,n}$; and restoration time is a random variable β_i with the distribution function $G_i(t) = P(\beta_i \le t), i = \overline{0,n}$. Preventive maintenance is carried out according to the age strategy [2]. Maintenance is carried out after operation for τ_i. It completely restores the elements. Maintenance duration is a random variable β_i^p with the distribution function $G_i^p(t) = P(\beta_i^p \le t)$, $i = \overline{0,n}$.

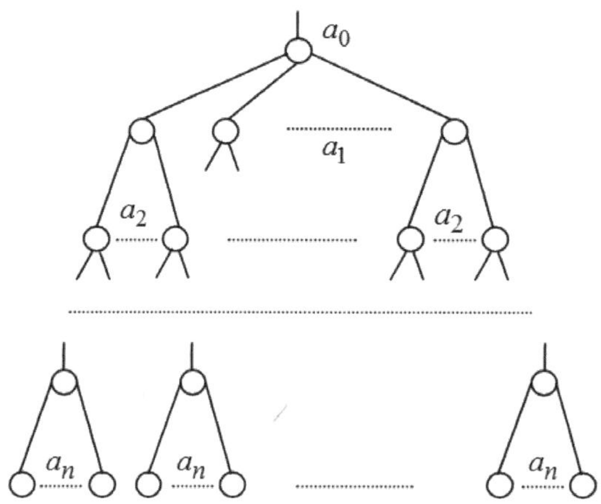

Figure 1. Structural scheme of resource-supplying network

The criterion of network failure is the absence of at least one operating way connecting the main and output elements.

The economical parameters of elements of ith layer ($i = \overline{0,n}$) of the network are: c_i is the income per unit of operation time, c_i^0 is the cost per unit of restoration time and c_i^p is the cost per unit of maintenance time.

The purpose of the research is to construct iteration algorithm of calculating stationary availability factor $K(\tau_0, \tau_1, ...\tau_n)$ of the network, average specific income $S(\tau_0, \tau_1, ..., \tau_n)$ per calendar time unit and average specific cost $C(\tau_0, \tau_1, ..., \tau_n)$ per time unit of network operation and to obtain optimal maintenance period of its elements.

2.1. Stationary characteristics definition

Let us denote by $K_i(\tau_i)$ the stationary availability factor, by $S_i(\tau_i)$ the average specific income, by $C_i(\tau_i)$ the average specific cost of the element of the ith layer. These characteristics are defined by the formulas (1) [2, 3]:

$$K_i(\tau_i) = \frac{T_i^{(1)}(\tau_i)}{T_i^{(1)}(\tau_i) + T_i^{(0)}(\tau_i) + T_i^{(2)}(\tau_i)};$$

$$S_i(\tau_i) = \frac{c_i\, T_i^{(1)}(\tau_i) - c_i^0\, T_i^{(0)}(\tau_i) - c_i^p T_i^{(2)}(\tau_i)}{T_i^{(1)}(\tau_i) + T_i^{(0)}(\tau_i) + T_i^{(2)}(\tau_i)}; \qquad (1)$$

$$C_i(\tau_i) = \frac{c_i^0\, T_i^{(0)}(\tau_i) + c_i^p T_i^{(2)}(\tau_i)}{T_i^{(1)}(\tau_i)},$$

where $T_i^{(1)}(\tau_i) = \int_0^{\tau_i} \overline{F}_i(t)dt$ is the average operation time; $T_i^{(0)}(\tau_i) = F_i(\tau_i)E\beta_i$ is the average restoration time; $T_i^{(2)}(\tau_i) = \overline{F}_i(\tau_i)E\beta_i^p$ the average maintenance time of the element of ith layer per regeneration period, $i = \overline{0,n}$.

To get stationary characteristics of the network we apply formulas from [3] to the structure considered. The following recurrent formulas are obtained. For the family of nth layer with a_n output elements controlled by one element of $(n-1)^{\text{th}}$ layer the characteristics are as follows:

$$K^{(n)}(\tau_n) = 1 - [1 - K_n(\tau_n)]^{a_n};$$

$$S^{(n)}(\tau_n) = a_n\, S_n(\tau_n);$$

$$C^{(n)}(\tau_n) = \frac{a_n\, C_n(\tau_n)\, K_n(\tau_n)}{K^{(n)}(\tau_n)}.$$

For one family of elements of mth layer having a_m elements controlled by one element of $(m-1)^{\text{th}}$ layer, the formulas are (for $m = \overline{1, n-1}$):

$$K^{(m)}(\tau_m,\ldots,\tau_n) = 1 - \left[1 - \frac{K_m(\tau_m)K^{(m+1)}(\tau_{m+1},\ldots,\tau_n)}{K_m(\tau_m) + K^{(m+1)}(\tau_{m+1},\ldots,\tau_n) - K_m(\tau_m)K^{(m+1)}(\tau_{m+1},\ldots,\tau_n)}\right]^{a_m};$$

$$S^{(m)}(\tau_m,\ldots,\tau_n) = \frac{a_m\left[S_m(\tau_m)K^{(m+1)}(\tau_{m+1},\ldots,\tau_n) + K_m(\tau_m)S^{(m+1)}(\tau_{m+1},\ldots,\tau_n)\right]}{K_m(\tau_m) + K^{(m+1)}(\tau_{m+1},\ldots,\tau_n) - K_m(\tau_m)K^{(m+1)}(\tau_{m+1},\ldots,\tau_n)};$$

$$C^{(m)}(\tau_m,\ldots,\tau_n) = \frac{a_m\left[C_m + C^{(m+1)}\right]\left[K_m + K^{(m+1)} - K_m K^{(m+1)}\right]^{a_m - 1}}{\sum\limits_{i=0}^{a_m - 1}\left[K_m + K^{(m+1)} - 2K_m K^{(m+1)}\right]^i \left[K_m + K^{(m+1)} - K_m K^{(m+1)}\right]^{a_m - i - 1}}.$$

The characteristics of the whole network with branching structure are:

$$K(\tau_0, \tau_1,\ldots\tau_n) = \frac{K_0(\tau_0)K^{(1)}(\tau_1,\ldots,\tau_n)}{K_0(\tau_0) + K^{(1)}(\tau_1,\ldots,\tau_n) - K_0(\tau_0)K^{(1)}(\tau_1,\ldots,\tau_n)};$$

$$S(\tau_0, \tau_1,\ldots,\tau_n) = \frac{S_0(\tau_0)K^{(1)}(\tau_1,\ldots,\tau_n) + K_0(\tau_0)S^{(1)}(\tau_1,\ldots,\tau_n)}{K_0(\tau_0) + K^{(1)}(\tau_1,\ldots,\tau_n) - K_0(\tau_0)K^{(1)}(\tau_1,\ldots,\tau_n)}; \tag{2}$$

$$C(\tau_0, \tau_1,\ldots,\tau_n) = C_0(\tau_0) + \frac{a_1\left[C_1 + C^{(2)}\right]\left[K_1 + K^{(2)} - K_1 K^{(2)}\right]^{a_1 - 1}}{\sum\limits_{i=0}^{a_1 - 1}\left[K_1 + K^{(2)} - 2K_1 K^{(2)}\right]^i \left[K_1 + K^{(2)} - K_1 K^{(2)}\right]^{a_1 - i - 1}}.$$

2.2. Optimization of maintenance period of network elements

Stationary characteristics of the network are the functions of ages τ_i of the elements. That is why the problem of optimal maintenance period for the elements is reduced to the problem of getting points of absolute extrema of the chosen criteria functions (2):

$$K(\tau_0, \tau_1,\ldots\tau_n) \to \max_{\tau_i \in (0,\infty);\, i = \overline{0,n}};$$

$$S(\tau_0, \tau_1,\ldots\tau_n) \to \max_{\tau_i \in (0,\infty);\, i = \overline{0,n}};$$

$$C(\tau_0, \tau_1, \ldots \tau_n) \to \min_{\tau_i \in (0,\infty);\, i=\overline{0,n}} .$$

2.2.1 Example 1: electricity supply network. As a rule, electricity travels big distances from electric stations to the customers. So, to organize electricity supply a branching network including elements changing technical characteristics of electricity are necessary [4, 5].

Let us consider main elements of the electricity supply network: 2 transformers increasing the voltage (elements of the first layer), and 36 transformers gradually decreasing the voltage (6 transformers up to 220 kV and 30 transformers to 6-10 kV). According to the instructions, maintenance of the network elements has to be carried out not less than once $\tau_0 = 100\, days,\ \tau_1 = 80\, days,\ \tau_2 = 60\, days,\ \tau_3 = 30\, days$ to ensure the necessary reliability of the network. The random variables $\alpha_i,\ \beta_i,\ \beta_i^P$ for the network elements have Erlangian distributions with the distribution functions:

$$F_i(t) = 1 - e^{-\lambda_i t} \sum_{j=0}^{2} \frac{(\lambda_i t)^j}{j!}, \quad G_i(t) = 1 - e^{-\mu_i t} \sum_{j=0}^{2} \frac{(\mu_i t)^j}{j!}, \quad G_i^P(t) = 1 - e^{-\mu_i^P t} \sum_{j=0}^{3} \frac{(\mu_i^P t)^j}{j!}.$$

Input characteristics of the elements are given in the table 1.

Table 1. Input data for the network from example 1

Layer number	Amount of family elements	Average time-to-failure $M\alpha_i$, *days*	Average restoration time $M\beta_i$, *days*	Average maintenance time $M\beta_i^P$, *hours*	Element's income c_i, *m.u./month*	Element's cost per restoration c_i^0, *m.u./month*	Element's cost per maintenance c_i^P, *m.u./month*
0	$a_0 = 1$	200	8,6	16,0	1500	2000	500
1	$a_1 = 2$	75	6,0	14,5	1200	1600	300
2	$a_2 = 3$	54	4,6	14,1	1000	1200	200
3	$a_3 = 5$	46	3,8	12,9	1000	900	200

Table 2. Optimization results for different criteria (example 1)

Layer number	τ_i, *days*	τ_i^S, *days*	$S^{\max}$, *m.u./month*	S^*, *m.u./month*	τ_i^C, *days*	$C^{\min}$, *m.u./month*	C^*, *m.u./month*
0	100,0	79,6			44,2		
1	80,0	33,4			8,6		
2	60,0	26,2	34061,5	32882,1	9,8	623,5	1261,6
3	30,0	20,0			12,2		

In table 2 we denote by S^*, C^* economical indexes of the network in case of recommended strategy $\tau_0 = 100\, days,\ \tau_1 = 80\, days,\ \tau_2 = 60\, days,\ \tau_3 = 30\, days.$

Preventive maintenance at the levels of time-to-failure $\tau_i^S,\ \tau_i^C, i = \overline{0,3}$, improves the indexes for 3,5% and 50,6% correspondingly depending on the criterion.

2.2.2. Example 2: network equipment. Let us consider the example of calculating characteristics and maintenance period optimization for elements of the commutation network equipment at the enterprise.

Let us consider the network consisting of the central commutation and 6 commutations (elements of the first layer). Each commutation serves 15 personal computers. According to the instructions, maintenance of elements should be carried out not less than $\tau_0 = 215\,days$, $\tau_1 = 115\,days$, $\tau_2 = 70\,days$ to ensure necessary reliability of the network. The distribution functions of the random variables α_i, β_i, β_i^p describing network elements are:

$$F_i(t) = 1 - e^{-\lambda_i t} \sum_{j=0}^{3} \frac{(\lambda_i t)^j}{j!}, \quad G_i(t) = 1 - e^{-\mu_i t} \sum_{j=0}^{2} \frac{(\mu_i t)^j}{j!}, \quad G_i^p(t) = 1 - e^{-\mu_i^p t} \sum_{j=0}^{4} \frac{(\mu_i^p t)^j}{j!}.$$

Input characteristics of the elements are given in the table 3.

Table 3. Input data for the network from example 2

Layer number	Amount of family elements	Average time-to-failure $M\alpha_i$, *days*	Average restoration time $M\beta_i$, *days*	Average maintenance time $M\beta_i^p$, *hours*	Element's income c_i, *m.u./month*	Element's cost per restoration c_i^0, *m.u./month*	Element's cost per maintenance c_i^p, *m.u./month*
0	$a_0 = 1$	450	4,0	12,3	1500	2800	1300
1	$a_1 = 6$	225	3,5	11,6	1200	2400	1200
2	$a_2 = 15$	129	3,2	11,4	1000	2000	800

Table 4. Optimization results for different criteria (example 2)

Layer number	τ_i, *days*	τ_i^S, *days*	$S^{\max}$, *m.u./month*	S^*, *m.u./month*	τ_i^C, *days*	$C^{\min}$, *m.u./month*	C^*, *m.u./month*
0	300,0	214,3			159,7		
1	200,0	107,8	94782,56	93574,47	80,6	1405,2	2320,26
2	100,0	53,2			45,7		

In table 4 we denote by S^*, C^* economical indexes of the network in case of recommended strategy $\tau_0 = 215\,days$, $\tau_1 = 115\,days$, $\tau_2 = 70\,days$. Preventive maintenance at the levels of time-to-failure τ_i^S, $\tau_i^C, i = \overline{0,3}$, improves the indexes for 1,2% and 39,4% correspondingly depending on the criterion.

3. Conclusions

Iteration processes for calculating stationary reliability and economical indexes of networks with branching structure is constructed with regard to preventive maintenance of the elements. The examples of specific resource-supplying networks show the optimal choice of maintenance intervals can result in the network characteristics improvement. Stationary characteristics of the network can be improved in comparison with the existing structure: for electric supplying network the average specific income increases for 3,5%, the average specific cost reduces for 50,6%; for the network equipment the average specific income increases for 1,2%, the average specific cost reduces for 39,4%.

Acknowledgments

The reported study was funded by RFBR and Samara Region according to the research project № 17-48-630410

IOP Conf. Series: Materials Science and Engineering **312** (2018) 012019 doi:10.1088/1757-899X/312/1/012019

References

[1] Cherkesov G N 2005 *Reliability of Hardware-software Complexes* (St. Petersburg: Peter).
[2] Beichelt F and Franken P 1983 *Zuverlässigkeit und Instanphaltung, Mathematische Methoden* (VEB Verlag Technik, Berlin).
[3] Peschansky A I 2009 *System technologies: Dnepropetrovsk* **61** 29.
[4] https://samelectrik.ru/kak-proisxodit-peredacha-i-raspredelenie-elektroenergii.html.
[5] http://electrik.info/main/school/892-kak-peredaetsya-elektroenergiya-ot-elektrostanciy-k-potrebitelyam.html.

Method of optimization onboard communication network

G A Platoshin[1], N I Selvesuk[1], M E Semenov[2], V M Novikov[1]

[1]State Research Institute of Aviation Systems, Moscow, 125319, Russia
[2]Voronezh State University, Voronezh, 394018, Russia

E-mail: gaplatoshin@2100.gosniias.ru

Abstract. In this article the optimization levels of onboard communication network (OCN) are proposed. We defined the basic parameters, which are necessary for the evaluation and comparison of modern OCN, we identified also a set of initial data for possible modeling of the OCN. We also proposed a mathematical technique for implementing the OCN optimization procedure. This technique is based on the principles and ideas of binary programming. It is shown that the binary programming technique allows to obtain an inherently optimal solution for the avionics tasks. An example of the proposed approach implementation to the problem of devices assignment in OCN is considered.

1. Introduction

At present days active research is being carried out in the field of improvement the design process of modern onboard networks built on the principle of distributed modular electronics (DME) [1]. Since architecture and avionics account 30% of the total cost of an aircraft the design of an optimal architecture is a key task [2]. Due to a non-optimal design of architecture of the aircraft's control system the optimal functioning of some processes may be reduced. In this way, in some critical cases such a non-optimal functioning of the control system may lead to the situation when some tasks significantly affecting the flight safety cannot be properly performed. The solution of such problems is an expensive task, so it makes the aircraft more expensive and less competitive.

The result of designing the architecture of the DME is the optimal (in the sense of some set of the target parameters) architecture of avionics systems for the aircraft. Modern avionics systems based on the DME architecture are characterized by the significant increase in functions as well as by the degree of system integration [3,4]. Also the developed avionics systems should be easier, safer and cheaper (from the service point of view) in comparison with the systems of previous generations. Ensure the implementation of these requirements can be done not only by developing the new hardware configuration devices of the system, but also by optimization of the avionics' systems architecture. The objectives of designing of avionics systems architecture usually conflict with each other, e.g., the increase in reliability may lead to an increase in the weight of the aircraft. In order to take into account possible alternatives to implement the different options and create an existing avionics architecture in the design it is advisable to use an approach based on the mathematical modeling and modern computer technologies.

2. Levels of architecture's optimization for OCN

After collecting the available data the architecture design process begins. The architecture design of the onboard information-processing network is carried out at three levels [5]. At the first level all the available requirements for the architecture of the onboard computing network are taken into account. First level contains:

1) Assigning devices. Distributes devices among installation sites. The initial data is a set of device types and a set of installation locations.
2) Assigning tasks. Provides an unambiguous assignment of tasks to devices. Tasks can be assigned to the device if the device has sufficient resources of each required type, and if the corresponding type of device may accommodate this type of task.
3) Assignment of peripheral wires. It is a laying of peripheral wires from equipment that is not related to the DME system, to DME devices inside the structure of aircraft installations.
4) Assignment of communication lines. Provides a minimum weight of communication lines. In addition to the structure, the initial data is a set of devices and communication lines, as well as the topology of their connection.
5) Assignment of signals (virtual links) to communication lines. The initial data is a set of transmitted signals, the topology of devices and communication lines, as information on the distribution of tasks.

At the second level the optimization of the onboard computing network is carried out and the device types are optimized (in accordance with the necessary resources) taking into account the data of the first level. First level contains:

1) Optimizing device types. It carries out the coordination of tasks and devices. Based on a set of potential device types, it is determined how many devices of what type should be used in which locations. During this process tasks are assigned to specific devices.
2) Network optimization. The process of synchronous determination of network topology, cable routes and signal transmission paths. The initial data is a set of all signals and their resources, performance and security restrictions, on the one hand. On the other hand it is the properties of communication lines and switches, for example, the bandwidth and the number of ports.

At the third level the optimized DME architecture is created. Creates the architecture of the DME based on the requirements of the systems and the aircraft. The calculation is based on a model of systems which includes all tasks, signals, as well as their resources and security requirements. Based on this data, a set of devices, locations and a connection network are defined in parallel with the assignment of all tasks and signals.

3. Mathematical technique for optimization of OCN

Integer programming is chosen as the main mathematical technique for optimization in the avionics problems. Integer programming is a discrete analog of linear programming (LP). Most of the problems listed above can be formalized in the form of problems of such a type. The main purpose of the linear programming is to find the vector of n variables x^{LP}, which minimizes the linear target function f [6]:

$$f^T x^{\mathrm{LP}} \tag{1}$$

taking into account that

$$Ax^{\mathrm{LP}} \leq b \tag{2}$$

$$x^{\mathrm{LP}} \in \mathbb{R}_+^n \tag{3}$$

An integer program (IP) is obtained by limiting x^{LP} up to integer values.

In the particular case $x^{\mathrm{IP}} = x^{\mathrm{BP}} \in \{0,1\}^n$ the problem of binary programming (BP) takes place. It is defined as minimization of

$$f^{\mathrm{T}} x^{\mathrm{BP}} \tag{4}$$

taking into account that

$$Ax^{\mathrm{BP}} \leq b, \tag{5}$$

$$x^{\mathrm{BP}} \in \{0,1\}^n \tag{6}$$

The BP problems have the same degree of algorithmic complexity and can be solved using the same methods as the IP tasks. The proposed approach is not unique, namely the numerous examples of the application of heuristic algorithms (various modifications of the genetic algorithm, the algorithm of the ant colony, etc.) are well known. However, these algorithms have significant drawbacks. The main drawback of heuristic algorithms is the lack of standard approaches. Each task requires a unique solver and the selection of such a solver can only be carried out by a highly qualified expert. Heuristic algorithms require the configuration of a large number of parameters and cannot use the information on the gradients, which reduces their capability to solve the classical optimization problems.

4. An example of the implementation of a mathematical apparatus

As an example let us consider the level of assignment of devices that are mentioned above [7]. In order to solve the problem of assigning the devices using the BP methods the device positioning scheme in the installation sites is represented as a binary vector

$$x = \left(\begin{matrix} \mathcal{D}_1 & \mathcal{D}_2 & \mathcal{D}_1 \mathcal{D}_2 & \cdots\cdots\cdots\cdots & \mathcal{D}_{|\mathcal{D}|} \\ \mathcal{J}_1 & \mathcal{J}_2 & \cdots\ \mathcal{J}_1 & \cdots\cdots\cdots\cdots & \mathcal{J}_{|\mathcal{J}|} \\ x_1 & x_2 & x_3\ \cdots & \cdots\cdots\cdots\cdots & x_{|x^{\mathcal{D}}|} \end{matrix}\right) \tag{7}$$

Each element of the vector x_i represents a possible arrangement of devices, namely a special installation site that has sufficient resources to accommodate one device of this type. The characters above x_i represent the combination of the device $\mathcal{D}_j$ and the installation location $\mathcal{J}_k$. The element of the vector x_i is equal to «1», if the device $\mathcal{D}_j$ is assigned to the installation site $\mathcal{J}_k$ and «0», if it is not assigned. The number of all possible distribution options for a single device $|x^{\mathcal{D}}|$ is determined by the length of the vector x, which is the dimension of the search space. However, this includes the solutions that do not assign a device to any installation site or assign it to several locations at once. A single-valued assignment is achieved by means of an additional equality (where $\mathcal{D}(i)$ is a device inscribed over x_i):

$$\sum_{i \in \{1,\ldots,|x^{\mathcal{D}}| \,|\, \mathcal{D}(i)=\mathcal{D}_j\}} x_i = 1 \tag{8}$$

for each device j. For all devices this expression is represented as a constraint matrix in the form of equality:

$$A_{single}^{eq} = \left(\begin{array}{cccccc|c} \mathcal{D}_1 & \cdots & \mathcal{D}_1 & \mathcal{D}_2 & \cdots & \mathcal{D}_{|\mathcal{D}|} & \\ \hline 1 & \cdots & 1 & 0 & \cdots & 0 & \mathcal{D}_1 \\ 0 & \cdots & 0 & 1 & \cdots & \vdots & \mathcal{D}_2 \\ \vdots & & & & \ddots & \vdots & \vdots \\ 0 & \cdots & \cdots & \cdots & \cdots & 1 & \mathcal{D}_{|\mathcal{D}|} \end{array}\right) \tag{9}$$

The right-hand side of the constraints is:

$$b_{single}^{eq} = \begin{pmatrix} 1 \\ \vdots \\ 1 \end{pmatrix} \begin{matrix} \mathcal{D}_1 \\ \vdots \\ \mathcal{D}_{|\mathcal{D}|} \end{matrix} \tag{10}$$

In addition, assignments that exceed the resource limits for any installation location are unacceptable. This is expressed in the form of inequality:

$$\sum_{i=1}^{|x^{\mathcal{D}}|} \rho_{\Phi_j}(i)x_i \leq \rho_{\Phi_j}^{\mathcal{I}_k} \tag{11}$$

for each type of resource $\mathcal{J}_j$, which is available on site $\mathcal{I}_k$, where $\rho_{\Phi_j}(i)$ is the amount of site resources that are involved in a corresponding destination i, $\rho_{\Phi_j}^{\mathcal{I}_k}$ is a number of resources, $\mathcal{J}_j$ is a type of resources, which installation site is $\mathcal{I}_k$.

Constraints on the installation site should be taken into account before the optimization process begins. Undesirable combinations of devices and combinations of installation locations are deleted from the solution vector, so the search space narrows.

5. Optimization parameters for OCN architecture

The final architecture must be guaranteed to meet all the necessary requirements.

The main parameters of optimization of architecture are [8-13]:

1) Delivery guarantees and messages delivery time. For each type of aircraft, there are message sets that are important to deliver as quickly as possible. The loss or delay of such messages can lead to disaster.
2) The number of resources. Increase in number of different types of devices, increases the development costs, the initial provision, the administrative costs as well as the maintenance costs.
3) The mass of devices in the architecture. This parameter has a great influence on the fuel consumption, and on the efficiency of the aircraft, consequently.
4) The cost of architecture. The aircraft manufacturer expends these funds on equipment for the creation of an aircraft.
5) Restriction on the energy consumption. The more energy is consumed by the avionics system, the less energy remains for other purposes, or more energy needs to be generated, which increases the fuel consumption.
6) The space required by the avionics system on the board. This space is occupied by devices and cables in the frame of DME system. The maximum amount is limited.
7) Reliability and redundancy [14].

6. Conclusion

The main difficulties encountered by the developer in the layout of the DME system are:

- Resource constraints.
- Restrictions by the safety.
- Bandwidth limitations.
- Limit by size, etc.

Also, during the optimization of the architecture the additional conflicting goals lead to additional complexity. A large number of functions and hardware components create a variety of possible architectures that are difficult to estimate manually. For example, for the distribution of 20 functions between the five devices, there are several million combinations. Real architectures have 1000 functions distributed among 40 devices. At first, all the evaluated architectures should be checked for compliance with restrictions that reach several thousand. In this way the subset of the permissible architectures should be placed inside the set of architectures with the inherently performed restrictions. And finally, a large number of goals from different areas (partially contradicting each other) makes it difficult to choose the final architecture. The approach discussed in this article is based on the BP model which helps to keep track of the targets at any change in the architecture. Advantages of the proposed approach are:

- obtaining the optimal solution;
- avoiding the probability of unpredictable behavior (by full search of options);
- a large number of ready-made standard software products for which it is necessary to change the posing of the problem;
- software versatility (the problem being solved is described separately from the solution algorithm).

Acknowledgments
This work is supported by Russian Foundation for Basic Research (18-08-00463-a).

References

[1] Kucheryavy A A 2016 *Avionic (*St. Petersburg: Lan') p. 452
[2] Collinson R P G 2011 *Introduction to Avionics Systems* (Netherlands: Springer) p. 530
[3] Wolfig R and Jakovljevic M 2008 *Digital Avionics Systems Conference DASC IEEE/AIAA* 1.E.4–1 –1.E.4–10
[4] Kravtsov V 2009 *European and Russian Joint Avionics Forum – New Generation of IMA Solutions for Future Aircraft*
[5] Annighöfer B and Thielecke F 2015 *CEAS Aeronaut. J. T.* **6** (3) 485–496
[6] Mittelmann H 2007 22nd *Euorpean Conference on Operational Research*
[7] Bondarenko I B 2013 *NB Cybernetics and programming* **1** 16–20
[8] Matveev M G 2017 *Vestnik VSU System analysis and information technologies* **3** 26-34
[9] Pedar A and Sarma V 1983 *Computers, IEEE Transactions* **C-32** (10) 911–922
[10] Koeppen C 2006 *Methodik zur modellbasierten Prognose von Flugzeugsystemparametern im Vorentwurf von Verkehrsflugzeugen* (Aachen: TUHH schriftenreihe Flugzeug-Systemtechnik) p. 269
[11] Neumann K 2008 *Deutscher Luft- und Raumfahrtkongress DGLR*
[12] Novikov V M 2016 *Scientific readings on aviation dedicated to the memory of N.E. Zhukovsky* 292–300
[13] Platoshin G A 2017 *Actual problems of research in avionics: theory, maintenance, development: Sat. sci. Art. on materials IV All-Russian NPK AVIATOR* 230–233
[14] Rehage D 2009 *Zustandsmodellierung und zuverlässigkeitsanalyse fehlertoleranter systemarchitekturen auf basis von integrierter modularer avionic* (Hamburg-Harburg: TUFF) p. 220

IOP Conf. Series: Materials Science and Engineering **312** (2018) 012021 doi:10.1088/1757-899X/312/1/012021

Development of robotic mobile platform with the universal chassis system

I Ryadchikov[1], E Nikulchev[2]*, S Sechenev[1], M Drobotenko[1], A Svidlov[1], P Volkodav[1] and A Feshin[1]

[1]Kuban State University, Krasnodar, 350040, Russia
[2]Moscow Technological Institute, Moscow, 119334, Russia

*Email: nikulchev@mail.ru

Abstract. The problem of stabilizing the position of mobile devices is extremely relevant at the modern level of technology development. This includes the problem of stabilizing aircraft and stabilizing the pitching of ships. In the laboratory of robotics and mechatronics of the Kuban State University, a robot is developed. The robot has additional internal degrees of freedom, responsible for compensating for deflections - the dynamic stabilization system.

1. Introduction

The problem of stabilizing the position of mobile devices is extremely relevant at the modern level of technology development. This includes the problem of stabilizing aircraft and stabilizing the pitching of ships, stabilizing and improving the safety of motor vehicles.

Particular attention is paid to the problem of stabilization in the construction of robotic mobile means and exoskeletons for people with disabilities [1]. At the moment, there are no exoskeletons that would allow people with inactive legs to maintain a dynamic balance without using additional means of support.

Also, the problem of moving by step is not solved for a wide class of off-road devices using the principle of walking for moving. According to the leading US universities and companies, expressed in the report "A Roadmap for US Robotics 2016": "In order to optimize the automation of the logistics chain processes in the world, robots must have the mobility that corresponds to the human level. Robots should be able to overcome stairs, escalators, doorways, curbs, concrete fragments, unpredictable environments, and move just like people. This type of extended mobility is possible for robots equipped with legs. Then the logistics will be 24/7 and inexpensive." From the same analysis: "Service robots that contain multiple mobility mechanisms, such as legs, caterpillars and high-speed wheels, will be able to provide collision-free and mobile manipulation in completely unstructured environments." According to this analysis, the market of mobile robots in need of stabilization will amount to 16 billion dollars by 2025. Jun Ho Oh, professor of engineering mechanics at the Korea Institute of Advanced Technology (KAIST, winner of the DARPA Robotics Challenge): "There is no right (for a walking robot) even for one mistake. The problem of stabilization of walking robots is not solved "[2]. Seth Teller of the Massachusetts Institute of Technology (MIT) team argues that the most difficult tasks for robots are in a rapidly changing environment. To solve these problems, it is necessary to search for the optimal configuration of mechanics [3].

The next problem is the movement in a highly rugged terrain in conditions of low or absent gravity. Existing rovers working on Mars and on the Moon cannot overcome a highly crossed, stony, rocky and hilly terrain, now an astronaut is required for such tasks [4]. Also, for work on the ISS, manipulators are attached to the station itself, access to the elements of the construction is limited to the zone of their operation. Currently, the task is to create mobile robotic systems capable of performing various operations inside and around the ISS [4].

The most relevant, based on the analyzed sources, is the creation of a platform for service robots with the characteristics described in the Terms of Reference for this project, namely:

The calculated speed of the platform moving with the dimensions of the body, fitting in the sphere of 40 cm, along the horizontal, on the average surface, in the step mode will be 5 km / h, and in the rolling mode on the body with the manipulators up to 15 km / h, it can autonomously function for 24 hours in the video broadcast mode, under active control and moving through the terrain medium cross-over, able to carry 3.8 kg payload.

As a result of the analysis of performance indicators, power-torque characteristics, market availability and prices, it was revealed that there is an actual problem with the selection of servo drives for walking structures with a carrying capacity of up to 5 kg, the weight of the structure up to 20 kg, and the battery life of up to 2 hours.

Thus, there is a wide class of devices that need more and more advanced stabilization technologies.

A two-legged walking robot can be developed by a specialist in practice and, when developed, ensures the realization of the claimed destination. The possibility of development in practice follows from the fact that for each feature included in the utility model formula on the basis of the description, the material equivalent is known, which makes it possible to conclude that it follows the criterion "industrial applicability" for the utility model and the "completeness of disclosure" criterion for the utility model.

In accordance with the proposed utility model, the applicant made a prototype of the Bipedal Walking Robot. The experimental operation of the proposed device showed that the static stability zones look very simple and account for up to 250% of the horizontal dimensions of the robot body and up to 100% of the length of the horizontally elongated support. In this case, the robot can overcome obstacles that exceed its own dimensions.

All this, in the final analysis, ensures the fulfillment of the achieved technical result – increasing the patency and overcoming obstacles.

The proposed device can be used as a walking robot:

- for educational purposes to study the appropriate modes of displacement;
- in robotics for moving devices;
- as an independent vehicle that allows you to move on a complex surface, for example, over the mountains, as well as on the surface of other planets.

The two-legged walking robot works as follows. (An example of use that does not limit the use of the utility model is given) [5, 6].

2. Mathematical modeling of the walking platform in development with the kinematic scheme

Mathematical modeling of the developing stepping platform with a kinematic scheme was performed to determine the force-moment characteristics of the drives used to ensure the movement of the joints of the robot's extremities. The mathematical modeling of the dynamic stabilization system was performed separately. Modeling was performed in the MATLAB matrix laboratory using the Simulink extension and specialized utility packages, such as the Control Toolbox, Simscape and CONTACT FORCES Library, to calculate the contact forces of interaction with the underlying surface. The general view of the model is shown in Figure 1. The Task subsystem translates the motion pattern, which is an array of specified angles of moving servo drives from the MATLAB data space. The displacement array is formed by the motion planner, and at the moment consists of empirically selected angles and angular velocities of the servo drives, allowing for a steady motion in increments.

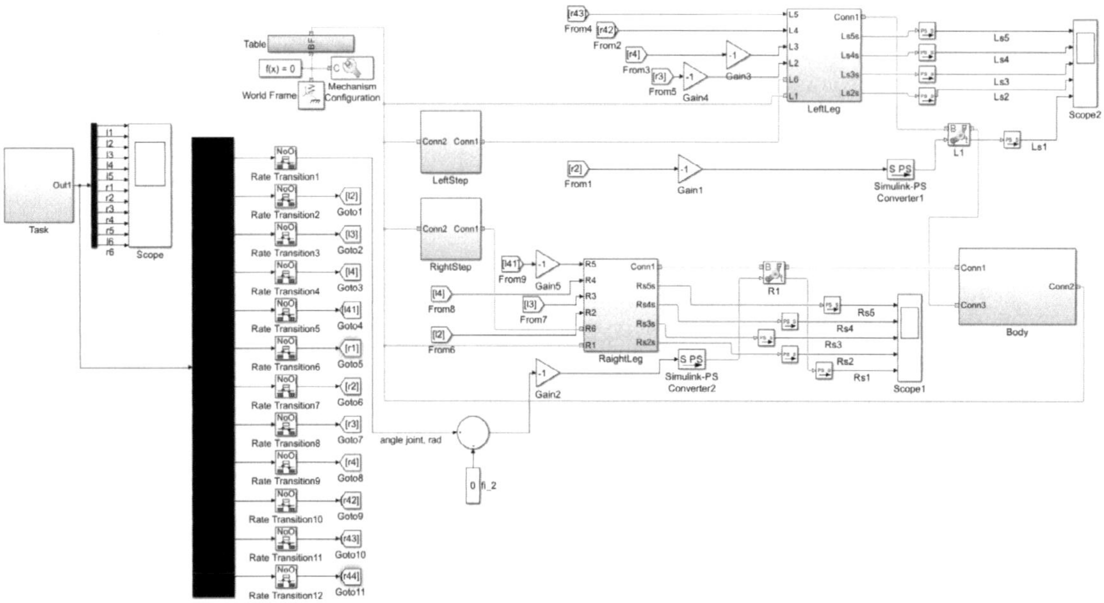

Figure 1. General diagram of the mathematical model in the MATULAB/Simulink package

3. Selection of servo drives of the robot propulsor

In the process of analyzing the kinematic scheme of the robot and numerical modeling of the walking process, the necessary servo drive parameters were identified, which are listed in Table 1.

Table 1. Found parameters of the required servo

Torque, Nm	Speed, rad / s	Weight, g	Feedback
30	1,5	300	Availability

In the process of analyzing the servo drives on the market, according to the resources http://www.servodatabase.com/, http://www.robotshop.com/, https://www.pololu.com/, http: //www.trossenrobotics .com, https://www.servocity.com/, www.robodrive.com, http://www.maxonmotor.com, http://hitecrcd.com, http://www.oc-servo.com, http://harmonicdrive.de servo drives were selected, suitable for any parameter.

The servo drive HS-1100WP Ultra-Heavy Duty, Giant, Waterproof Servo was chosen (Fig. 2).

Figure 2. Servo Drive HS-1100WP

This servo drive has a waterproof, extremely rugged body and high torque, which allows the drive to be used on an industrial scale and in bad weather conditions. At the same time, the HS-1100WP is very compact and light, just 0.363 kg. However, the big drawback of this drive is the lack of feedback.

4. Conclusion

The proposed robotic system is able to move in an environment adapted to humans, having a non-anthropomorphic kinematic scheme. In this case, for the movement and stabilization, there are different cyberphysical systems, which made it possible to develop a mathematical model of controlled motion with an analytically solvable system of equations.

Advantages of the kinematic scheme AnyWalker over anthropomorphic constructions (for example, solutions from Boston Dynamics):

- low center of gravity and many zones of static stability;
- large surgical field with small dimensions (the robot in height of 70 sm can reach to a 2-meter obstacle);
- energy efficiency and low value of the Cost of Motion parameter;
- scalability;
- possibility to climb on obstacles, 2.7 times exceeding the dimensions of the platform.

The principles of independence of kinematic walking and stabilization are supposed to be used in developing a solution for stabilizing exoskeletons, which will allow the pilot to accelerate movement and get rid of crutches.

In the future, the system of dynamic stabilization consisting of three flywheels realized in the form of motor-wheels or driven by gears each from its motor located in orthogonal planes will be developed (while the mass centers of flywheels coincide). One flywheel is perpendicular to the ground and the direction of motion, the second is perpendicular to the ground and is parallel to the direction of motion, and the third is parallel to the ground. This stabilization system allows compensating for external influences without changing the kinematics of walking, as well as the moments of inertia that arise when moving the manipulators.

Acknowledgments

The work was carried out within the framework of the state task of the Ministry of Education and Science, project No. 8.2321.2017 “Development and adaptation of control systems for compensation of dynamic deflecting effects on mobile objects in a state of dynamic equilibrium”.

References

[1] Rupal B S, Rafique S, Singla A et al 2017 *International Journal of Advanced Robotic Systems* **14** (http://journals.sagepub.com/doi/pdf/10.1177/1729881417743554)
[2] Guizzo E and Ackerman E 2015 *IEEE Spectrum* **52** 11.
[3] Keisner A, Raffo J and Wunsch-Vincent S 2016 Robotics: Breakthrough Technologies, Innovation, Intellectual Property. *Forsait* **2** [In Rus]
[4] Computing Community Consortium 2016 A roadmap for US robotics: From Internet to robotics (http://jacobsschool.ucsd.edu/contextualrobotics/docs/rm3-final-rs.pdf)
[5] Ryadchikov I, Sechenev S, Svidlov A, Sinitsa S and Nikulchev E 2017 *MATEC Web of Conferences* **99** 02007
[6] Ryadchykov I, Sechenev S et al 2017 *International Journal of Advanced Computer Science and Applications* **8**(9) 29

IOP Conf. Series: Materials Science and Engineering **312** (2018) 012022 doi:10.1088/1757-899X/312/1/012022

Airborne data measurement system errors reduction through state estimation and control optimization

G G Sebryakov[1], S M Muzhichek[1], V I Pavlov[2], O V Ermolin[2] and A A Skrinnikov[1 *]

[1]State Research Institute of Aviation Systems, Moscow, 125319, Russia
[2]Tambov State Technical University, Tambov, Russia
[3]Central Recearch Institute of the Air Force of the Ministry of Defense of the Russian Federation, Moscow region, Russia

*E-mail: msm19@yandex.ru

Abstract. The paper discusses the problem of airborne data measurement system errors reduction through state estimation and control optimization. The approaches are proposed based on the methods of experiment design and the theory of systems with random abrupt structure variation. The paper considers various control criteria as applied to an aircraft data measurement system. The physics of criteria is explained, the mathematical description and the sequence of steps for each criterion application is shown. The formula is given for airborne data measurement system state vector posterior estimation based for systems with structure variations.

1. Introduction.

Airborne data measurement systems (DMS) are widely used in aviation. Irrespective of DMS type, sensor spectral band, design features, etc., one of major vehicle DMS requirements is the obtaining epy reliable information on current situation, as well as the timeliness of information.

Airborne DMS has the following parameters changing in a wide range: alternating vibration g-loads, temperature, humidity and pressure due to various environmental conditions of vehicle operation. In addition, DMS is susceptible to interference in the spectral range of its sensors and to jamming.

The paper proposes approaches to improve onboard DMS functioning for two cases: a fault-free properly functioning DMS and a system with errors. The causes of these errors can be both natural – ageing and wear – and intentional – jamming. DMS state recognition in abovementioned conditions is a very urgent problem in radar, navigation and communication areas. Besides, it is advisable to solve this problem without normal system operation interruption.

Let us consider the solution to the problem by the example of an airborne radar system.

2. Statement of problem

We consider an active airborne radar with passive response and an airborne digital computer as parts of an aircraft DMS. The aircraft task is to locate air vehicles, probably equipped with radar jammers. Because of ageing, wear and other adverse factors there might be gradual negative changes in DMS characteristics and its component features. The airborne radar may receive not only useful return signals but also deception signals, generated by jammers. It is essential to recognize the aircraft DMS state, specifically, to find out whether DMS functions in a routine mode or in conditions of abnormal measurements due DMS faults or jamming. DMS functioning in regimes of abnormal measurements is associated with possible changes of system structure.

3. Method of system state recognition in inaccurate measurements

The joint aircraft and airborne radar state n-dimensional **X**-vector characterizes center-of-mass motion kinematics, the dynamics of motion relative to the center-of-mass, radar and target relative positioning. It also includes airborne radar basic states such as angular misalignment φ between aircraft axis and target direction; line of sight angular rate ω, aircraft velocity relative to the target V and distance to the target D.

In addition, aircraft constructive element bending vibrations, environmental parameters, etc. can be taken into account.

Random inputs and outputs are typical for airborne radars. Let's consider an airborne radar observation model. In this case, a generalized measurement model can be presented in the form of m-dimensional vector $\boldsymbol{Z}$ and random measurement noise vector $\boldsymbol{N}$

$$\boldsymbol{Z}(t)=\boldsymbol{C}\boldsymbol{X}(t)+\boldsymbol{D}(t)\boldsymbol{N}(t)$$

where $\boldsymbol{C}$ - measurement matrix; $\boldsymbol{N}(t)$ – m-dimensional centered Gaussian white measurement noise vector with correlation function $\boldsymbol{K}_N(t,t_1)=\boldsymbol{Q}(t)\delta(t-t_1)$; $\boldsymbol{Q}(t)$ – measurement noise intensity matrix; $\delta(t-t_1)$ – Dirac function, $\boldsymbol{D}(t)$- noise amplification matrix. In cases of failures and abnormal measurements caused by other reasons the matrix $\boldsymbol{D}(t)>>\boldsymbol{I}$; for uninformative measurements measurement matrix C – zero matrix, $\boldsymbol{D}(t)$ – unity matrix; in deception jamming environment vector $\boldsymbol{X}(t)$ is influenced by the jamming signals.

The generalized object model for joint aircraft and airborne radar state vector we use in the form of a variable structure model [1]

$$\boldsymbol{X}_{k+1}=\boldsymbol{\Phi}_{k,\,k+1}^{(s)}\boldsymbol{X}_k+\mathbf{B}_k^{(s)}\boldsymbol{U}_k+\mathbf{F}_k^{(s)}\xi_k, \tag{1}$$

where $\boldsymbol{X}_k$ - joint aircraft and airborne radar state vector; $\boldsymbol{\Phi}_{k,\,k+1}^{(s)}$ – aircraft and radar state transition matrix; k – discrete current time; s – index, corresponding to the aircraft and radar structure number, active in k time moment; $\mathbf{F}_k^{(s)}$ - specified matrices which components are the functions of measurement vector; $\boldsymbol{U}_k$ – aircraft control vector relative to a selected target; ξ_k - centered discrete Gaussian noise vector with correlation function matrix $\mathbf{K}_\xi(k,h)=\mathbf{G}_k\delta_{kh}$, h – time moment different from k; G_k – aircraft state noise intensity matrix; δ_{kh} - Kronecker delta.

For variable structure airborne radar measurements can be written in the form [3]

$$\mathbf{Z}_k=\mathbf{C}_k^{(s)}(\boldsymbol{\mu}_k,\boldsymbol{\gamma}_k)\boldsymbol{X}_k+\mathbf{N}_k^{(s)}\zeta_k; \tag{2}$$

$$\boldsymbol{\mu}_k=f_{k-1}^{(s)}(\boldsymbol{\mu}_{k-1},\boldsymbol{\gamma}_k),\qquad \mu_0=\hat{\mu}_0,\quad k=\overline{1,\,K}, \tag{3}$$

where $\boldsymbol{Z}_k$ –measurement vector; $\mathbf{C}_k^{(s)}(\boldsymbol{\mu}_k,\boldsymbol{\gamma}_k)$ - non-random measurement matrix that is a function of $\boldsymbol{\mu}_k$ и $\boldsymbol{\gamma}_k$, parameters, defining measurement conditions in s-th structure; $\boldsymbol{\mu}_k$ – measured parameters set matrix; $\boldsymbol{\gamma}_k$ - measurement process control vector with limitations

$$\boldsymbol{\gamma}_k\in\Gamma_K,\quad g(\boldsymbol{\mu}_k)\le\overline{g}\,. \tag{4}$$

In (2)-(4) function $f_{k-1}^{(s)}(\cdot)$, $g(\boldsymbol{\mu}_k)$, $\overline{g}$ value and Γ_k set – are specified initial parameters; $\mathbf{N}_k^{(s)}$ – specified matrix; ζ_k – measurement noise vector; i.e. centered discrete Gaussian noise vector with correlation function matrix $\mathbf{K}_\zeta(k,\,h)=\mathbf{Q}_k\delta_{kh}$, $\boldsymbol{Q}_k$ – measurement noise intensity matrix; μ_0, $\hat{\mu}_0$ -

respectively, measured parameters and their estimates at the initial moment of normal system operation.

In countermeasures environment measurement vector errors might be taken into account using measurement control procedure, in this case radar measurement vector components are formally presented in the form

$$\varphi + \Delta\varphi; \quad \boldsymbol{\omega} + \Delta\boldsymbol{\omega}; \quad \boldsymbol{V} + \Delta\boldsymbol{V}; \quad \boldsymbol{D} + \Delta\boldsymbol{D},$$

that is

$$\boldsymbol{X} = \boldsymbol{X} + \Delta\boldsymbol{X}. \tag{5}$$

It is assumed that $\Delta\boldsymbol{X}$ components gradually change from zero to maximum values, which are individual for every parameter. Let's consider measurement control problems for specific control types $\{\gamma_k\}$ [3].

4. Measurement control problems. Measurement program (mode) selection

In measurement program selection γ_k and μ_k are scalar parameters. Γ_k set consists of two elements: $\Gamma_k = \{0, 1\}$, we assume $\gamma_k = 1$, if at the moment k the measurement is made; $\gamma_k = 0$, if there is no measurement. Then equation (3) for s-th structure can be written as

$$\mu_k = \mu_{k-1} + \gamma_k, \quad \mu_0 = 0$$

with limitations

$$\mu_K = \sum_{k=1}^{K} \gamma_k \le K_\Sigma,$$

where K_Σ – a specified number of measurements.

In equation (2)

$$\mathbf{C}_k^{(s)}(\boldsymbol{\mu}_k, \gamma_k) = \gamma_k \mathbf{C}_k^{(s)},$$

where $\mathbf{C}_k^{(s)}$ – sensitivity of measurement channel in s-th structure.

5. Measured parameter combination selection

When selecting parameters for measurement in the equation (3) we formally write down $\mu_k = \gamma_k$, and measurement matrix in (2)

$$\mathbf{C}_k^{(s)}(\boldsymbol{\mu}_k, \boldsymbol{\gamma}_k) = \mu_k.$$

Thus, matrix control $\boldsymbol{\gamma}_k \in \Gamma_K$ specifies the combination of measured parameters, and Γ_k set is a potentially possible parameter combination.

6. Onboard DMS position (flight path) selection

In some cases, we may enhance measurement system efficiency through better operating conditions. For example, we may form the aircraft flight path in order to reduce airborne radar measurement errors.

In such a case the model (1) – (3) can be written as:

$$\boldsymbol{X}_{k+1} = \boldsymbol{\Phi}_{k,k+1}^{(s)} \boldsymbol{X}_k + \mathbf{B}_k^{(s)}(\boldsymbol{X}_k)\boldsymbol{U}_k + \mathbf{F}_k^{(s)}\xi_k; \tag{6}$$

$$\boldsymbol{Z}_k = \mathbf{C}_k^{(s)}(\boldsymbol{X}_k, \boldsymbol{U}_k)\boldsymbol{X}_k + \mathbf{N}_k^{(s)}\zeta_k; \tag{7}$$

$$\mathbf{C}_k^{(s)}(\boldsymbol{X}_k, \boldsymbol{U}_k) = C_{0k}^{(s)}(\boldsymbol{U}_k) + \sum_{j=1}^{n} c_{jk}^{(s)}(\boldsymbol{U}_k)\boldsymbol{X}_k, \tag{8}$$

where $C_{0k}^{(s)}(\,\boldsymbol{U}_k)$ - nonlinear functions; $c_{jk}^{(s)}$ – statistical linearization coefficients for centered state variables. $C_{0k}^{(s)}$ and $c_{jk}^{(s)}$ are calculated according algorithms [4] with the use of Gaussian approximation of posterior probability density [5] and depend on $\boldsymbol{U}_k$ control inputs, posterior mathematical expectations $\hat{\boldsymbol{X}}_{jk}^{(s)}$ and noise intensity matrices $\mathbf{G}_k$ and $\mathbf{Q}_k$.

The additional component in the model (6) – (8) is the control optimization procedure $\boldsymbol{U}_k$ in accordance with a pre-validated criterion. For example, if the measurement control purpose is just filtering quality improvement, then we shall consider average losses as an "information" criterion

$$\boldsymbol{I}(\hat{\boldsymbol{X}}_k) = M\left[\Psi(\boldsymbol{X}_k,\, \hat{\boldsymbol{X}}_k)\right] = \sum_{s=1}^{S}\int_{-\infty}^{\infty}\Psi\left(\boldsymbol{X}_k,\, \hat{\boldsymbol{X}}_k\right)\hat{p}_k(\boldsymbol{X})\,d\boldsymbol{X}; \tag{9}$$

$$\Psi\left(\boldsymbol{X}_k,\, \hat{\boldsymbol{X}}_k\right) = \sum_{j=1}^{n}\left[\boldsymbol{X}_{jk} - \hat{\boldsymbol{X}}_{jk}\right]^2, \tag{10}$$

where $M[\cdot]$ – mathematical expectation operation; $\Psi\left(\boldsymbol{X}_k,\, \hat{\boldsymbol{X}}_k\right)$ – quadratic loss function; $\hat{p}_k(\boldsymbol{X})$ – posterior probability density of state vector $\boldsymbol{X}$.

If the purpose of measurement control is the simultaneous filtering quality enhancement and airborne measurement characteristics improvement through flight path optimization, then the problem of aircraft conrol should be solved according to a generalized "information and geometric factor" criterion

$$\boldsymbol{J}\left(\hat{\boldsymbol{X}}_k\right) = \alpha\boldsymbol{I}\left(\hat{\mathbf{X}}_k\right) + \beta\boldsymbol{L}\left(\hat{\boldsymbol{X}}_k\right); \tag{11}$$

$$\boldsymbol{L}\left(\hat{\boldsymbol{X}}_k\right) = \left[\hat{\boldsymbol{X}}_k - \boldsymbol{X}_{\text{опт}}\right]^2, \tag{12}$$

where α and β –coefficients weighting the requirements for $\boldsymbol{X}$ vector filtering accuracy and optimal positioning of airborne radar relative to a selected target. With that $\boldsymbol{I}\left(\hat{\boldsymbol{X}}_k\right)$ is determined according to (9), (10); $\boldsymbol{X}_{\text{opt}}$ – aircraft state vector providing optimal measurement conditions for a selected target.

The solution of this problem necessitates aircraft position control in accordance with the rule

$$\boldsymbol{U}_k = \begin{cases} \boldsymbol{U}_{\max}, \text{ при } \boldsymbol{U}_k \geq \boldsymbol{U}_{\max}; \\ \boldsymbol{U}_{\text{o. y}}, \text{ при } \boldsymbol{U}_k < \left|\boldsymbol{U}_{\max}\right|; \\ -\boldsymbol{U}_{\max}, \text{ при } \boldsymbol{U}_k \leq -\boldsymbol{U}_{\max}, \end{cases}$$

where $\boldsymbol{U}_{\text{o. y}}$ is a control input determined according to the methodology [6], ensuring that airborne radar location is a compromise between two goals reflected in the criterion (11). $\boldsymbol{U}_{\max}$ – control input maximum value.

For simultaneous stability in the intersensor interference environment and the required filtering quality an airborne radar optimization should be carried out in accordance with the following criterion

$$\boldsymbol{F}\left(\hat{\boldsymbol{X}}_k\right) = \boldsymbol{I}\left(\hat{\boldsymbol{X}}_k\right) + M\left(\hat{\boldsymbol{X}}_k, \boldsymbol{Z}_k,\, \boldsymbol{U}_k\right); \tag{13}$$

$$M\left(\hat{\boldsymbol{X}}_k, \boldsymbol{Z}_k, \boldsymbol{U}_k\right) = \int_{-\infty}^{\infty}\Psi(\boldsymbol{X}_k,\, \hat{\boldsymbol{X}}_k)\hat{p}_k(\boldsymbol{X})\,d\boldsymbol{X}\int_{-\infty}^{\infty}\phi_k^{(s)}(\boldsymbol{X},\boldsymbol{Z},\boldsymbol{U})\hat{p}_k^{(s)}(\boldsymbol{X})\,d\boldsymbol{X}; \tag{14}$$

$$\phi_k^{(s)}(\boldsymbol{X}, \boldsymbol{Z},\, \boldsymbol{U}) = \sum_{l,\,q=1}^{n}\frac{\bar{Q}_{lq}^{(s)}}{\left|\mathbf{Q}^{(s)}\right|}\left[\boldsymbol{Z}_k - \mathbf{C}_{lk}^{(s)}(\mu_k,\, \gamma_k)\boldsymbol{X}_k\right]\left[\boldsymbol{Z}_k - \mathbf{C}_{qk}^{(s)}(\mu_k,\, \gamma_k)\boldsymbol{X}_k\right], \tag{15}$$

where l, q – indices of airborne radar state vector components; $\bar{Q}_{lq}^{(s)}$ – algebraic complement of element Q_{lq} in the determinant $\left|\mathbf{Q}^{(s)}\right|$ of measurement system noise matrix. In this case, the control becomes two-level [7]. The first-level control is carried out according to the criterion (14) for each s-th structure, and results in the selection of the best s number. At the second level, the problem of filtering quality improvement is solved in accordance with the criterion (9).

The estimation of a DMS state vector is based on the posterior probability density $\hat{p}_k^{(s)}(\boldsymbol{X})$ of an aircraft state vector in the s-th structure which is determined by the Bayes's rule based on the prior probability density $p_k^{(s)}(\boldsymbol{X})$ and $\boldsymbol{Z}_k$ measurement:

$$\hat{p}_k^{(s)}(\boldsymbol{X}) = \frac{p_k^{(s)}(\boldsymbol{X})\exp\left[-0{,}5\,\phi_k^{(s)}(\boldsymbol{X},\ \boldsymbol{Z},\ \boldsymbol{U})\right]}{\sum\limits_{s=1}^{S}\int\limits_{-\infty}^{\infty} p_k^{(s)}(\boldsymbol{X})\exp\left[-0{,}5\phi_k^{(s)}(\boldsymbol{X},\ \boldsymbol{Z},\ \boldsymbol{U})\right]d\boldsymbol{X}}, \tag{16}$$

7. Conclusion

Thus, the problem of reduction an airborne DMS errors is discussed for two mostly typical cases:; operational flight with properly functioning system and DMS use in jamming environment. In the first case the problem is solved traditionally based on the criterion (9), the aircraft flight path optimization may also be introduced in accordance with (12) and minimizing the criterion (11); in the second case – it is necessary to determine possible jamming effects according to the criterion (14) and with that to provide the required filtering quality in accordance with the criterion (13). The problem solution is based on the joined aircraft-DMS state vector $\{\boldsymbol{X}_k,\ s_k\}$ posterior probability estimation using the theory of systems with random abrupt structure variation [7].

References

[1] Pavlov V, Aksyonov V and Belova T 2010 Measurement systems optimization. *Proc. of Tomsk Polytechnic University* **317**(4) 104-06.
[2] Ventzel E and Ovcharov L 2013 *Theory of random processes and its engineering applications. Teaching guide.* (Moscow. KNORUS)
[3] Malyshev V, Krasilschikov M, and Karlov V 1989 *Air vehicle navigation and control optimization.* (Moscow. Mashinostroyeniye)
[4] Gusev A 2009 *Probabilistic methods in vehicles and machinery.* (Moscow: BMSTU)
[5] Ledoux M, and Talagrand M 1991 *Probability on Banach Spaces.* (New York: Springer-Verlag)
[6] Khoroshavin V, Zotov A and Ishutinov D 2014 *Mechatronics, automation, control* **2** 23
[7] Bukhalev V 1996 *Recognition, assessment and control in systems with random abrupt structure variation.* (Moscow: Nauka Physmatlit)

IOP Conf. Series: Materials Science and Engineering **312** (2018) 012023 doi:10.1088/1757-899X/312/1/012023

The heat storage material based on paraffin-modified multilayer carbon nanotubes with Nickel-zinc ferrite

A Shchegolkov[1], A Shchegolkov[1], T Dyachkova[1] and A Semenov[2, 3 *]

[1] Tambov State Technical University, Tambov, 392000, Russia
[2] Institute of Space Technologies, People's Friendship University of Russia (RUDN University), Moscow, 17198, Russia
[3] Moscow Technological Institute, Moscow, 119334, Russia

*E-mail: semenov.venture@mail.ru

Abstract The paper presents an investigation of magnetically controlled heat-storage material based on paraffin, modified with multilayer carbon nanotubes with nickel-zinc ferrite. The technology of obtaining nanomodified material capable of interacting with magnetic field is presented. The study of the heat-exchange processes of charge/ discharge with the help of magnetic field are carried out.

1. Introduction.

The development of modern materials science is associated with the emergence of technologies for the production of "smart" materials possessing special functional properties and capable of adapting to various environmental conditions. Heat-storage materials may take on special importance in the hierarchy of creating "smart" materials. Heat-storage materials can be widely used in aerospace engineering, power engineering, as well as for heat-loaded elements of power and computer electronics. For aviation important is reliability and minimum mass-dimensional parameters. As the directions of application of heat-storage materials, it is possible to isolate electronic power components, microclimate thermal control systems, electric energy storage systems, especially if they are built on the basis of lithium-ion and polymer batteries. The ability to absorb thermal energy for such materials is a consequence of the accumulation of heat as a result of a phase transition. As a rule, these are phase transitions in the range from 40 to 120 ° C. Recently, heat-storage materials are actively used as the basic elements of thermal interfaces. Such thermal interfaces can be used to thermoregulate lithium-ion and lithium-polymer batteries. This concerns both portable systems and powerful storage stations for renewable energy and electric transport. This is the direction of scientific research. It was proposed in [1] to use the material to change the thermophysical parameters of paraffin by using graphene. The aim of the work was to improve the performance of lithium-ion batteries due to the reduction of the negative effect of self-heating. In [2], the effectiveness of the discharge efficiency of a lithium cell at various operating temperatures (15-45 ° C) was estimated by the thermal model. Numerical results were compared with experimental data obtained from laboratory studies. In [3], a comparison was made between the internal and external cooling procedures for thermal control. They performed a two-dimensional and three-dimensional transient temperature analysis of a prismatic lithium-ion battery. Consider the prospects for using nanomaterials for phase transition materials using the example of theoretical work [4, 5]. In [4], to improve the properties of paraffin, a theoretical substantiation of nanocrystals of graphene, garnet nanoplates, and nanomembranes h-BN is carried out. Thermal conductivity of paraffin with 1% and 4% vol. Is considered. Concentrations of these nanofillers. In this case, the thermal conductivity of paraffin with fillers was 0.3 W / mK and 0.6 W / mK (0.36 W / mK and 0.65 W / mK). The paper [5] shows that the use of graphene in the phase-transition material makes it possible to increase its thermal conductivity by more than two orders of magnitude while retaining the latent ability to store heat. The use of this approach leads to a significant drop in temperature in a standard lithium-ion battery.

The paper [6] shows studies of the thermal conductivity of materials with a self-leveling effect in the structure of a modified material. A method for the synthesis of nanocomposite graphene with Fe3O4 nanoparticles is presented. Self-leveling of "magnetic graphene" improves the thermal conductivity of composites and epoxy resins. Improvement of thermal conductivity with oriented fillers occurs at low concentrations of the order of ~ 1%. In real time, testing with computer chips demonstrated a temperature decrease of 10 ° C using a material without a thermal interface with ~ 1 wt. % oriented fillers. At the same time, the authors of [7] consider materials that do not have a phase transition, which reduces their effectiveness in the case of intense heat fluxes. Questions arise with giving nanostructures magnetic properties. The functionalization of CNTs with the help of magnetic nanoparticles was investigated in Refs [7,8]. The influence of orientation and alignment of carbon nanotubes (CNTs) embedded in an epoxy polymer matrix under the action of a magnetic field on the mechanical properties of the nanocomposite obtained was studied in [7,8]. The authors found that when alignment of CNTs occurs in fact in high magnetic fields, the properties of the composites obtained exceed properties that were not subjected to a magnetic field. So in [7,8] physicochemical mechanisms are investigated under which positive charges ensure effective adsorption of negatively charged magnetic nanoparticles on the surface of graphene by means of electrostatic interactions. The adsorption of nanoparticles (diameter D ~ 6-10 nm) on graphene surfaces is more effective than on CNT surfaces due to the high curvature of CNTs, which prevents the formation of dense coatings. Magnetic nanoparticles obtained in solution (basic pH) are negatively charged and therefore are electrostatically attracted to the positively charged layer of poly-dimethyl-diallylammonium chloride adsorbed on graphene fillers. For the CNT, the authors of Ref. [7] established that the pH for the most effective adsorption of Fe3O4-ɣ-Fe2O3 nanoparticles on polyelectrolyte was found to be pH = 11.9-12.0 [8]. Hybrid materials based on maghemite (γ-Fe2O3) / multilayer carbon nanotubes (MWCNT) were synthesized in [9] and their anisotropic electrical conductivities were investigated as a result of their equalization in a polymer matrix under an external magnetic field. Modification of γ-Fe2O3 nanoparticles on the MWCNT surface was achieved by a modified sol-gel reaction in which sodium dodecylbenzenesulfonate (NaDDBS) was used to inhibit the formation of a 3D iron oxide gel. These hybrid materials, in particular magnetized multi-walled carbon nanotubes (m-MWCNT), were easily aligned parallel to the direction of the magnetic field even when a weak magnetic field was used. However, such methods of functionalization were considered in the framework of applied problems aimed at changing the strength and electrical conductivity of polymers. In this connection, it is necessary to consider the possibility of obtaining efficient heat-storage materials with improved thermophysical properties and adapted for use in aeronautical engineering. Heat-accumulating materials for aviation equipment should have functionality by using additional materials that react to the magnetic field. The possibility of interaction with a magnetic field, which will acquire nanomodified heat-accumulating material, will provide an opportunity to regulate internal heat exchange. This will allow you to control the charge and discharge modes of the thermal battery, which will be based on magnetically controlled heat-accumulating material, which will make the material more compact and provide a reduction in mass-dimensional parameters. In addition, the problem of moving layers of high-temperature layers with a lower temperature regime of convective heat transfer will be considered.

2. Methodology and materials

To implement the tasks set, the technology of directed synthesis of multilayer carbon nanotubes was used. The introduction of MWNTs (multilayer carbon nanotubes) into paraffin was carried out with the help of ultrasonic radiation and exposure to magnetic fields (up to 1500 mT). Within the framework of directed production of carbon nanotubes, the synthesis of Mo-Al_2O_3-MgO and Ni-MgO catalysts was carried out. After that, MWNTs were mechanically activated with nickel-zinc ferrites 400NH Ni1-xZnxFe_2O_4 with preliminary crushing to 2-4 microns in a ball mill for 60 minutes. Subsequently, the resulting carbon nanotubes were used to modify paraffin P-2 (Lukoil, Russia).
The modification of paraffins obtained by (multilayer carbon nanotubes) with the necessary morphology provides interactions at the molecular level and creates a stable material with the characteristics that are necessary for the formation of TM. The modification occurs during the phase transition of paraffin, when it passes into the liquid state at the temperature at which the threshold of its thermal decomposition is reached, which allows the reactivity of MWNTs with paraffin molecules to initiate this process. The application of ultrasonic radiation (2 kW power) made it possible to create cavitation in the volume of liquid paraffin (at a temperature of 90-95 ° C).

To ensure the necessary orientation of the structures integrated in the MWNTs, the molten paraffin was exposed to a magnetic field during cooling. After that, the material was placed in the extruder using a fluoroplastic screw. The power of the drive motor of the extruder is 2 kW at a speed of 100 rpm. In addition, the resulting material was granulated in a compacting granulator to form granules to a size in the range of 0.5 to 1 mm. After this operation, the granules were placed in a carrier fluid. The carrier fluid was a synthetic motor oil CASTROL 0W30. Motor oil and magnetic filler were weighed on the Unit G-200 scales with an accuracy of ± 5 mg. After this, the filler was introduced into the engine oil and mixed thoroughly. The resulting thick slurry was diluted with a carrier liquid and mechanically stirred until a colloidal solution was obtained. As a result, the liquid phase - motor oil was 20%, and the solid phase - nanomodified composite - 80%.

Figure 1 shows an experimental model for the study of charge / discharge in magnetically controlled heat-accumulating material.

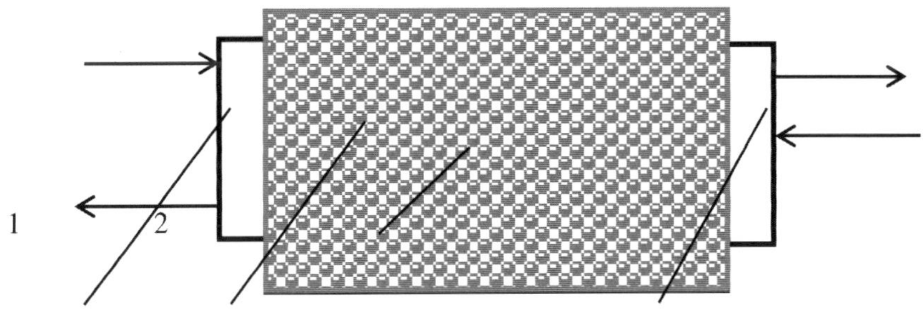

Where 1 - input heat exchanger, 2 - vessel c in magnetically controlled heat-accumulating material distributed in industrial oil, 3 - magnets 4 - output heat exchanger

Figure 1. Experimental model for charge / discharge research in magnetically controlled heat-accumulating material

The heat flux source is located in the lower part of the tank in the form of a flat electric heater with a power of 300 W and an area of 0.003 m^2. With such heater parameters, the heat flow is 100 kW / m^2

Research results Preliminary test for interaction with a magnetic field (Fig. 2) Fig. 2 A shows the heat-accumulating composite before the action of the magnetic field, and in Fig. 2 B after exposure.

Figure 3 shows the distribution of the temperature field, taking into account the effect of the magnetic field at the end part of the heating battery.

Figure 4 shows the distribution of the temperature field with allowance for the effect of further movement of the magnetic field.

Figure 2. Ferromagnetic liquid under the action of a magnetic field.

Figure 3. 3-dimensional thermogram of magnetically controlled heat-storage material

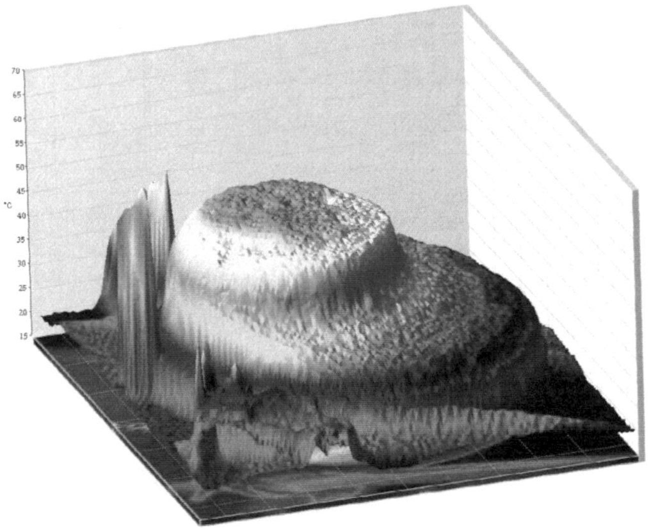

Figure 4. 3-dimensional thermogram of magnetically controlled heat-storage material.

Figure 5 shows the distribution of the temperature field with allowance for the effect of the magnetic field over time - 10 s.

Figure 6 shows the distribution of the temperature field taking into account the effect of the magnetic field over a period of 25 s.

Figure 7 shows the distribution of the temperature field taking into account the effect of the magnetic field over a period of 45 s.

Figure 5. 3-dimensional thermogram of magnetically controlled heat-storage material

Figure 6. 3-dimensional thermogram of the magnetically controlled heat-storage material.

Figure 7. 3-dimensional thermogram of magnetically controlled heat-storage material

3. Discussion of the results

Figure 3-7 shows the distribution of the temperature field in the container containing the nanomodified paraffin with ferromagnetic particles distributed in the synthetic oil, as a result of the action of the magnetic field of the cylindrical shape created by the permanent magnet. To intensify the process, a permanent magnet is used, which creates a magnetic field in a specific part of the container and causes TM to travel to this part of the container. In the case of mechanical displacement of the magnet, the intensity of heat exchange increases, since it becomes possible to transfer hot and cold layers. The subsequent equalization of the temperature field from the state (Fig. 6) with an explicit temperature peak is due to intrinsic thermal conductivity (Fig. 7). At the end of the charge of the heating battery, it is possible to create voids in front of the heat exchangers, which increases the efficiency of storing thermal energy.

It should be taken into account that the orientation of the magnetic moments in the direction of the applied field is impeded by thermal motion. Taking into account both factors, as in the classical theory of paramagnetism of Langevin, leads to a formula for the magnetization of a magnetic fluid:

$$M = M_0 L\left(\frac{mH}{kT}\right) \quad (1)$$

$$M_0 = nm = \varphi M_s, L(\xi) = \mathrm{cth}\xi - \xi^{-1} \quad (2)$$

Because of the large magnitude of the moment m, the nonlinear effects appear quite early: the value of ξ 1 is reached at room temperature even in fields H ~ 102. If in conventional suspensions the gradients of the particle concentration are due solely to Archimedean forces, then for magnetic suspensions placed in an inhomogeneous field H , a gradient of the magnetic field plays a role analogous to the gravitational field. In an inhomogeneous field, the force (m) H acts on a particle with a magnetic moment m. Estimating the magnitude of this force, it should be borne in mind that for those particle sizes (d ~ 100 A), which are used in stable magnetic colloids, each suspended particle is separate magnetic domain. Calculating the critical dimensions below which the particle becomes absolutely single-domain leads to values of d from several hundred angstroms (330 and 760 A for iron and nickel, respectively).

4. Conclusions

1. The mechanisms of heat exchange in magnetic fluids based on synthetic oil containing ferromagnetic particles (nickel-zinc ferritic), paraffin and multilayered carbon nanotubes are studied.
2. A technique for experimental studies for a magnetic fluid based on synthetic oil containing ferromagnetic particles (nickel-zinc ferrite), paraffin and MWNTs has been developed. The range of measurements of the temperature field was from 20 to 75 ° C.
3. Experimental investigations of the temperature field of a magnetic fluid under various temperature conditions under the influence of a magnetic field produced by neodymium magnets with a magnetic induction of 1500 mT, as well as their analysis.

Acknowledgments
The research was carried out within the framework of the grant of the Russian Science Foundation (project No. 15-13-10038).

References
[1] Goli P, Legedza S, Dhar A, Salgado R, Renteria J and Balandin A 2014 *Journal of Power Sources* **248** 37
[2] Kumaresan K, Sikha G and White R E 2008 *Journal of the Electrochemical Society* **2** 155
[3] Mohammadian S, He Y and Zhang Y 2015 *Journal of Power Sources* **293** 458
[4] Mortazavi B, Yang H, Mohebbi F, Cuniberti G and Rabczuk T 2017 *Applied Energy* **202** 323
[5] Goli P, Legedza S, Dhar A, Salgado R, Renteria J and Balandin A 2014 *Journal of Power Sources* **248** 37
[6] Renteria J et al 2015 *Materials and Design* 241-221
[7] Correa-Duarte M A, Grzelczak M, Salgueiriño-Maceira V et al 2005 *J. Phys. Chem.* **109** 19060
[8] Garmestani H, Al-Haik M, Dahmen S et al 2003 *Adv. Mater* **15** 1918
[9] Kim T, Tannenbaum A and Tannenbaum R 2011 *Carbon* **49** 54

IOP Conf. Series: Materials Science and Engineering **312** (2018) 012024 doi:10.1088/1757-899X/312/1/012024

Methods and tools for profiling and control of distributed systems

R Sukharev[1], O Lukyanchikov[1], E Nikulchev[1,2] * D Biryukov[2] and I Ryadchikov[3]

[1]Moscow Technological University (MIREA) Moscow, Russia
[2]Moscow Technological Institute, Moscow, Russia
[3]Kuban State University, Krasnodar, Russia

*E-mail: nikulchev@mail.ru

Abstract. This article is devoted to the topic of profiling and control of distributed systems. Distributed systems have a complex architecture, applications are distributed among various computing nodes, and many network operations are performed. Therefore, today it is important to develop methods and tools for profiling distributed systems. The article analyzes and standardizes methods for profiling distributed systems that focus on simulation to conduct experiments and build a graph model of the system. The theory of queueing networks is used for simulation modeling of distributed systems, receiving and processing user requests. To automate the above method of profiling distributed systems the software application was developed with a modular structure and similar to a SCADA-system.

1. Introduction

All complex systems tend to be separated into smaller modules that perform autonomous tasks. Each developed software module may be included in different programs, if the conditions of its use are met, declared in the documentation for this module. Thus, a software module can be considered as a tool to deal with the complexity of the programs, and as a tool to combat duplication in programming. But modular programming has failed to fully implement all its concepts in full [1]. Therefore, with the development of networks and distributed systems, modules became separate network applications, which formed a service-oriented system Software-as-a-Service (SaaS) and more complex cloud systems. However, the complexity of the analysis of such systems also increased, since operations are performed on different computing nodes, and for each service a different load is applied. So today it is important to develop methods and tools for profiling and management of distributed systems.

All existing applications for the analysis of distributed systems are based on the analysis of logs gathered from all services, such as ELK [2, 3] or Dapper [4] by Google. These tools allow finding errors or bottlenecks using the logs provided by the developers, but do not allow to effectively take corrective measures, that is, perform a control action on the system. Therefore, systems like SCADA (Supervisory Control And Data Acquisition) are preferred for profiling and management of distributed systems. The term "SCADA-system" is used when it comes to automated systems, that is, systems of control and management, carried out with the participation of a person – the dispatcher (operator). The information systems produce a great amount of operations. In result, the operator is unable to keep up, so it is preferable to only show false and time-consuming operations, this will help finding the bottlenecks in the system and reconfigure it if necessary.

2. Methods for Profiling of Distributed Systems

Today there is no widely accepted methods and tools for profiling of distributed systems, but there is a number of proposed methods in [5, 6]. Both of these methods for analysis and profiling of distributed systems are focused on simulation to conduct experiments and build the graph model of a system. After combining these methods, it is possible to identify the main stages of profiling distributed systems: (1)To form the graph representation of the process of solving functional problems as a coherent set of query processing. (2)To build and configure the simulation model of query processing for a specific version of a computer network and the distributed data warehouse. (3)To carry out the experiment on the simulation model to evaluate the execution time of the queries. (4)To determine the bottlenecks in the system, dramatically affecting the performance of the distributed system.

In these methods the profiling results depend on the simulation model, which determines the load on the system with a certain requests' flow intensity. When profiling a distributed system using the calculation indicators for monitoring the system's effectiveness, following should be considered:

- a plurality of storage devices in the queuing systems (QS). This factor leads to a consideration of the QS as a system with a total lineup of applications, eliminating the receipt of the individual characteristics of the queues;
- generally non-deterministic time for the service of application unit. The time spent on application maintenance depends on the total volume of the package, with the result that there is a mutual dependence of the average time of service applications and the number of requests waiting for service;
- correlations between the time of receipt of applications coming from different elementary sources, leading to lack of steady-state flow.

Having calculated performance of individual QS's, it is possible to calculate the performance of stochastic queuing network (SQN), defined by a set of characteristics described in [6]: (1) set $\{S_1, S_2, ..., S_n\}$, forming a network; (2) number of channels $K_1, K_2, .., K_n$ in the systems $S_1, S_2, ..., S_n$, respectively; (3) trajectory matrix of requests $R=||r_{ij}||$, where r_{ij} is the QS number, to which the request is addressed on the i-th path in the j-th service phase during the deterministic routing procedure; or matrix of the probabilities of the requests' transfer from one QS to another: $P=||p_{ij}||$, where p_{ij} is the probability that a request leaving S_i goes to S_j; (4) number of requests circulating in an isolated network (Z); (5) intensity of request sources in an open network $\Lambda=\{\lambda_i(k)\}$, where i is the type of the request, and k is the category of urgency; (6) time distribution laws $F_1(t)$, $F_2(t)$, ..., $F_n(t)$ and request maintenance disciplines in the systems $S_1, S_2, ..., S_n$. The systems $S_1, S_2, ..., S_n$ and the connections between them determine the network structure. The intensity of the request sources $\lambda_{ij}(k)$, the service intensity $\mu_{ij}(k)$, the queue lengths, and the operating mode of the devices characterize the load and performance of the SQN.

Modeling of distributed information based on SQN system described in [7–9].

After building the simulation model, for profiling of a distributed system it is necessary to conduct a series of experiments with different system architectures, in which the number of calculated nodes, their location, data flow, and other parameters change [10, 11]. SCADA-like software will allow automating the construction of these systems, immediate implementation and the collection of data based on the results of the experiment.

3. Development Tools for Control and Profiling

Tools that automate profiling of distributed systems are implemented in hmiSCADA system (https://github.com/RSukharev/hmiSCADA), written in C ++, using the Qt libraries. This application uses the principle of modular construction. The system consists of a main module, providing a complete set of basic functions for processing and visualization of information, as well as additional functional modules, implementing the data collection functions and optional monitoring and control functions in the application. Additional modules for data collection and management are included in the source code of the profiled system.

IOP Conf. Series: Materials Science and Engineering **312** (2018) 012024 doi:10.1088/1757-899X/312/1/012024

The core module's functionality includes: (1) obtaining data from external modules; (2) data visualization; (3) options to manage the software profiling process; (4) simulation.

Additional modules' functionality includes: (1) data collection on the times of start and end of the test piece of code; (2) sending profiling data over the network to the main module of the system; (3) execution of the commands, received from the visualization and control modules.

The main difference of this system from other SCADA systems is that to render data received from an operator, it is not required to build the mnemonic of what should be displayed in advance. The mnemonic is generated automatically based on the data obtained. However, due to the modular structure of the system it is possible in the future to bring the graphical user interface to a common standard. It is also possible to provide compatibility with the data transmission standards for SCADA systems (OPC UA). Enhanced system functionality occurs as new modules are written and connected.

Modules are connected dynamically and have following interfaces: (1) request, intended for sending data; (2) processData, designed to execute commands and process the data received.

The work of the system is based on the exchange of messages between modules. The messages are exchanged using a data structure consisting of fields: «senderID» (identifier of the sending plugin), «receiverID» (identifier of the receiving plugin), «command» (the command to execute on the side of the receiving plugin), and «value» (forwarded data).

The architecture of the system allows following the process, in which messages may pass through a series of modules, each of which performs its work on data processing and forwards the result further, modifying and extending the functionality of the system as a whole.

The components of the system are shown in Figure 1:

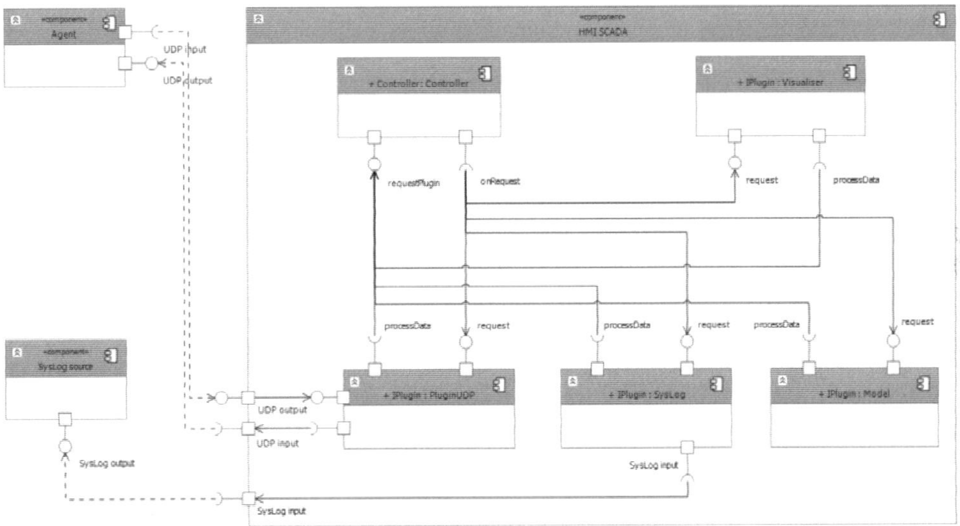

Figere 1. Diagram of components of the hmiSCADA system.

The main component in the diagram (hmiSCADA) consists of the following elements:

- Controller (message manager and provides data exchange between system modules);
- Visualiser (visualization module, provides a visualization of the received data as a mnemonic scheme and provides a graphical user interface for controlling the profiling process);
- PluginUDP (network module, provides the reception of messages over the network protocol UDP from the module AgentUDP and redirection to the Controller component, as well as sending over the network of control commands from the Controller to the AgentUDP module);

- Syslog (the log analysis module is used to obtain profiling data based on log analysis of the system);
- Model (module designed for simulation).
- Additional modules of the system are represented in the diagram by components:
- Agent (data acquisition and management module for the profiled system);
- SysLog source (any data source that is compatible with the Syslog format).

The collection of information is ensured by including in the source code a profiled system of the Profiler class and implemented through following macros:

```
define PROFILE_BEGIN(nodeinfo, moduleinfo)
         Profiler _profiler_(nodeinfo, moduleinfo, __FILE__);

#define PROFILE_START(...)
         _profiler_.start( __FILE__, __LINE__, ##__VA_ARGS__);

#define PROFILE_STOP()
          _profiler_.stop(profileBlock, __FILE__, __LINE__);

#define ON_EXIT(...) _profiler_.onExit();
```

Macros call functions of the Profiler class, with parameters: the name of the file with the source code and the line number of the profiled block.

Example of use:

```
#include "Profiler/profiler.h"

std::string appID = qPrintable(QString("mini") +
       QString::number(getTimeNS()));

       #define PROFILE_ENABLE

       #define PROFILE_BEGINNER PROFILE_BEGIN(appID, " ");
                                    PROFILE_START(Q_FUNC_INFO);
       #define PROFILE_ENDER PROFILE_STOP();

       void mini::on_pbSend_released() {
               PROFILE_BEGINNER;
               sendMessage();
               PROFILE_ENDER;
       }
```

Between PROFILE_BEGINNER and PROFILE_ENDER macros, an arbitrary number of lines of code can be placed, the main requirement is only the general execution context (within one function or within one block). The Agent module is the network and simultaneously managing the profiler module.

AgentUDP sends a message to the address and port of the main module of the system over the network. The recipient in the message is the Visualiser module. PluginUDP gives the data received via the network to the Controller class, adding its own identifier to it as a parameter «pluginID» (repeater plugin ID). Controller redirects the data to the module specified in the message as the recipient, i.e. Visualiser. In the message, the sender specifies the identity of the AgentUDP module, so if you need to send a response or a control command, the Visualiser module will generate a message with this ID as the recipient. The Visualiser message will be sent to the Controller message manager and then forwarded to the relay module (PluginUDP), which will forward it over the network to the AgentUDP module with the corresponding identifier.

The results of profiling a simple chat sending message by UDP are shown in Figure 2.

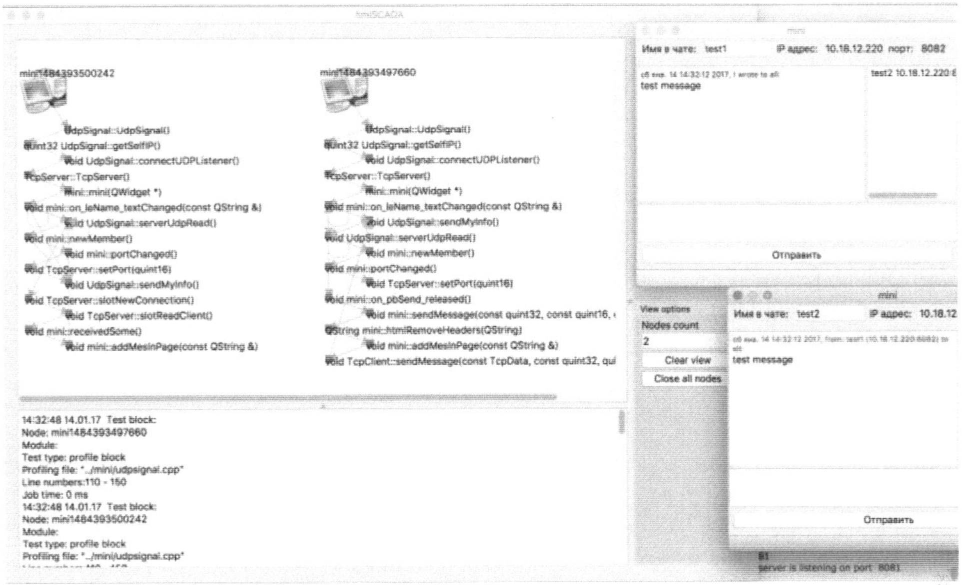

Figure 2. Visualization of profiling results of two instances of the chat application.

Instances of profiled applications (Figure 2) are marked with an icon showing the workstation. Functional blocks are marked with gray squares with the name of the profiled function. The sequence of transitions between the functional blocks is indicated by arrows. When you hover the cursor on the function block, a pop-up window appears with information about the execution time. Active functions and transitions between them are highlighted in green. The menu for controlling the instance of the profiled application is called up by right-clicking on the icon. By default, there are options for displaying information about the application and completing the application.

As a result, this SCADA-type profiler allows not only operating as a sniffer and show packets in the system, but also visually shows the messages passing in the system at the function level and notices the time of their execution, thus when messages are delayed or lost they quickly find problematic places. This visual approach to mapping processes in the system and the approach of modular integration into existing systems can solve the problems posed with the complexity of analysis and profiling of distributed systems.

4. Conclusion

This article describes the methods and SCADA-like tools for profiling and management of distributed systems. Those allow automating the process of profiling systems, which makes it possible to locate bottlenecks and errors in the system.

An important factor in the development and promotion of these SCADA-like systems is the ability to integrate them into functioning distributed systems. That's why the prototype hmiSCADA proposes modular architecture. With this architecture, third-party developers can develop their modules for hmiSCADA. Costs for the development of these modules are easily compensated by the time saved on profiling and analysis of the system.

Acknowledgments
The work was carried out within the framework of the state task of the Ministry of Education and Science, project No. 8.2321.2017 "Development and adaptation of control systems for compensation of dynamic deflecting effects on mobile objects in a state of dynamic equilibrium

References

[1] Scott W R and Davis G F 2015 *Organizations and organizing: Rational, natural and open systems perspectives* (Routledge)
[2] Reelsen A 2014 Using elasticsearch, logstash and kibana to create realtime dashboards Available: https://secure.trifork.com/dl/goto-berlin-2014/GOTO_Night/logstash-kibana-intro.pdf, 2014.
[3] Bagnasco S et al 2015 *Journal of Physics: Conference Series* **608** 012016.
[4] Sigelman B H et al 2010 *Dapper, a large-scale distributed systems tracing infrastructure* (Technical report, Google)
[5] Lukyanchikov O, Filatov V et al 2016 *ITM Web of Conferences* **6** 03009
[6] Ivanov A Y, Gorshkov V S *Methods for assessing the effectiveness of distributed information stores* Available: http://vestnik.igps.ru/wp-content/uploads/V22/7.pdf [In Russian]
[7] Gurvich I, Van Mieghem J A 2014 *Manufacturing & Service Operations Management* **17** 16.
[8] Hershey P C, Rao S, Silio C B, Narayan A 2015 *IEEE Systems Journal* **9** 212.
[9] Bramson M. 2008 *Stability of queueing networks* (Springer Berlin Heidelberg).
[10] Nikulchev E, Pluzhnik E et al 2015 *International Journal of Advanced Computer Science and Applications* **6**(6) 22.
[11] Nikulchev E, Pluzhnik E et al 2016 *International Journal of Advanced Computer Science and Applications* **7**(1) 30.

IOP Conf. Series: Materials Science and Engineering **312** (2018) 012025 doi:10.1088/1757-899X/312/1/012025

Nonparametric method for failures diagnosis in the actuating subsystem of aircraft control system

M N Terentev, S S Karpenko, E Yu Zybin and V V Kosyanchuk

State Research Institute of Aviation Systems, Moscow, Russia

E-mail: eyzybin@2100.gosniias.ru

Abstract. In this paper we design a nonparametric method for failures diagnosis in the aircraft control system that uses the measurements of the control signals and the aircraft states only. It doesn't require a priori information of the aircraft model parameters, training or statistical calculations, and is based on analytical nonparametric one-step-ahead state prediction approach. This makes it possible to predict the behavior of unidentified and failure dynamic systems, to weaken the requirements to control signals, and to reduce the diagnostic time and problem complexity.

1. Problem setting for fault diagnosis

Let the models of a non-failure and failure aircraft dynamic be represented in the state space as

$$x_{i+1} = Ax_i + Bu_i , \tag{1}$$

$$x_{j+1}^f = Ax_j^f + B_f u_j , \tag{2}$$

where $i = \overline{0, l-1}$, $j = l, l+1, \ldots$ are the discrete times before and after the occurrence of failures; l is the instant a failures occur; x, x^f are the non-failure and failure aircraft state vectors of length n_x ; u is the control vector of length; A , B are the matrices of eigen-dynamics and control efficiency; $B_f = BF$ is the control efficiency matrix of the failure aircraft; $F = \operatorname{diag}\left[f(1)\ f(2) \cdots f(n_u) \right]$ is the matrix of failures, $f(*) = 1$ for the non-failure control channels, $0 \le f(*) < 1$ for the failure control channels. It is necessary, based on the measurements of control signals u and states x only, to diagnose the failures in aircraft control system by determining the residuals in aircraft control efficiency parameters $\Delta B = B - B_f$. These residuals can be very useful for subsequent failure compensation by means of a aircraft control system reconfiguration [1–2].

There are two different approaches to solution of this problem [3–8]. The first approach is based on identification of B and B_f directly from (1)–(2) with subsequent computation of ΔB . The main challenge of this approach is the unidentifiability of aircraft model parameters without special test signals and a priory information [9–11]. So, the simplest way to solve failures diagnosis problem is to use parametric one-step-ahead prediction approach with the help of known estimations $\hat{A}$, $\hat{B}$:

$$\hat{x}_{i+1} = \hat{A}x_i + \hat{B}u_i , \tag{3}$$

$$\hat{x}_{j+1}^{f} = \hat{A}x_{j}^{f} + \hat{B}u_{j}. \tag{4}$$

Then we can calculate the difference between real (2) and predicted (4) failure aircraft state values

$$\Delta x_{j+1}^{f} = x_{j+1}^{f} - \hat{x}_{j+1}^{f} = Ax_{j}^{f} - \hat{A}x_{j}^{f} + B_{f}u_{j} - \hat{B}u_{j} = \left(A-\hat{A}\right)x_{j}^{f} + \left(B_{f} - \hat{B}\right)u_{j}. \tag{5}$$

If the aircraft model parameters are known exactly, then $\hat{A} = A$, $\hat{B} = B$, $\Delta\hat{A} = 0$, $\Delta\hat{B} = \Delta B$, and ΔB due to (5) can be easily derived from simple equation

$$\Delta\hat{B}u_{j} = \Delta x_{j+1}^{f}. \tag{6}$$

The main problem of the parametric prediction methods is that they require a priori information about the model parameters, which may be uncertain or even completely undefined.

2. The solution of failures diagnosis problem by the nonparametric prediction method

Widely known nonparametric prediction methods (neural networks, cellular automata, Markovian, chaotic, etc.) require preliminary training/tuning for a particular system or based on statistical algorithms, which use a sufficiently large sample of measurement data for the statistical properties extraction [12–15]. Therefore, such methods are not applicable, for example, for maneuverable aircraft evolution prediction. However, the main disadvantage of the known non-parametric prediction methods is that they can't be easily applied to solve one-step-ahead state prediction problems for systems with failures. In this paper we propose to solve the problem by analytical nonparametric prediction method, which is not subject to model errors, do not use statistical calculations, do not require training, and can be used to predict unidentified and failure dynamic systems.

Let's write the one-step-ahead prediction expressions (3)–(4) in the matrix form

$$\hat{X}_{i+1} = \hat{A}X_{i} + \hat{B}U_{i},\ \hat{X}_{j+1}^{f} = \hat{A}X_{j}^{f} + \hat{B}U_{j}, \tag{7}$$

where $U_i = [u_i \ldots u_{i+h}]$, $U_j = \left[u_j \ldots u_{j+h^f}\right]$, $X_i = [x_i \ldots x_{i+h}]$, $X_j^f = \left[x_j^f \ldots x_{j+h^f}^f\right]$, $\hat{X}_{i+1} = [\hat{x}_{i+1} \ldots \hat{x}_{i+1+h}]$, $\hat{X}_{j+1}^f = \left[\hat{x}_{j+1}^f \ldots \hat{x}_{j+1+h^f}^f\right]$. Note that, in contrast to the modeling problems (1)–(2), the one-step-ahead prediction problems (7) can be written by the single seamless block matrix expression

$$\left[\hat{X}_{i+1}\ \hat{X}_{j+1}^{f}\right] = \left[\hat{A}\ \hat{B}\right]\begin{bmatrix} X_i & X_j^f \\ U_i & U_j \end{bmatrix}. \tag{8}$$

Let's rewrite (8) to the form of parameters estimation equation

$$\left[\hat{A}\ \hat{B}\right]\begin{bmatrix} X_k \\ U_k \end{bmatrix} = \hat{X}_{k+1}, \tag{9}$$

where $k = i$ before and $k = j$ after the occurrence of failures; $X_k = \left[x_k \ldots x_{k+h^*}\right]$, $\hat{X}_{k+1} = \left[\hat{x}_k \ldots \hat{x}_{k+h^*}\right]$, $U_k = \left[u_k \ldots u_{k+h^*}\right]$; $h^* = h$ for $k = 0 : l-1$ and $h^* = h^f$ for $k \geq l$.

It's known [2, 16, 17], that any linear matrix equation of the form $YC = D$ with known matrices C and D is solvable for Y if and only if the solvability condition holds

$$D\bar{C}^{R} = 0, \tag{10}$$

and all set of its solutions is defined by a formula

$$X = \begin{bmatrix} D\tilde{C}^R & \Theta \end{bmatrix} \begin{bmatrix} \tilde{C}^L \\ \bar{C}^L \end{bmatrix} = D\tilde{C} + \Theta \bar{C}^L , \tag{11}$$

where Θ is an arbitrary matrix; $\bar{C}^L$, $\bar{C}^R$ are the left and right zero divisors of maximal rank (matrices for which conditions $\bar{C}^L C = 0$, $C\bar{C}^R = 0$ are satisfied); $\tilde{C}^L$, $\tilde{C}^R$ are the left and right unity divisors ($\tilde{C}^L C \tilde{C}^R = I$); $\tilde{C} = \tilde{C}^R \tilde{C}^L$ is the generalized inverse matrix defined by a canonical decomposition

$$C = \begin{bmatrix} \tilde{C}^L \\ \bar{C}^L \end{bmatrix}^{-1} \begin{bmatrix} I & 0 \\ 0 & 0 \end{bmatrix} \begin{bmatrix} \tilde{C}^R & \bar{C}^R \end{bmatrix}^{-1} .$$

Subject to (10) the equation (9) is solvable if and only if

$$\hat{X}_{k+1} \overline{\begin{bmatrix} X_k \\ U_k \end{bmatrix}}^R = 0 . \tag{12}$$

The condition (12) is fulfilled for any moment of time and can be used to solve the one-step-ahead state prediction problem. Let's write the estimation equation (9) in form:

$$\begin{bmatrix} A & B \end{bmatrix} \begin{bmatrix} x_k & \dots & x_{k+h^*-1} & x_{k+h^*} \\ u_k & \dots & u_{k+h^*-1} & u_{k+h^*} \end{bmatrix} = \begin{bmatrix} \hat{x}_{k+1} & \dots & \hat{x}_{k+h^*} & \hat{x}_{k+h^*+1} \end{bmatrix} , \tag{13}$$

then the solvability condition (12) can be written as an expression

$$\begin{bmatrix} \hat{x}_{k+1} & \dots & \hat{x}_{k+h^*} & \hat{x}_{k+h^*+1} \end{bmatrix} \overline{\begin{bmatrix} x_k & \dots & x_{k+h^*-1} & x_{k+h^*} \\ u_k & \dots & u_{k+h^*-1} & u_{k+h^*} \end{bmatrix}}^R = 0 , \tag{14}$$

where h^* is the minimum number of observations for which the right zero divisor has the form of a column vector:

$$\begin{bmatrix} x_k & \dots & x_{k+h^*-1} & x_{k+h^*} \\ u_k & \dots & u_{k+h^*-1} & u_{k+h^*} \end{bmatrix} \begin{bmatrix} r_k(1) \\ \vdots \\ r_k(h^*) \\ r_k(h^*+1) \end{bmatrix} = 0 . \tag{15}$$

Hence the condition (14) takes the form

$$\begin{bmatrix} \hat{x}_{k+1} & \dots & \hat{x}_{k+h^*} & \hat{x}_{k+h^*+1} \end{bmatrix} \begin{bmatrix} r_k(1) \\ \vdots \\ r_k(h^*) \\ r_k(h^*+1) \end{bmatrix} = 0 , \tag{16}$$

that can be rewritten as an equation

$$\hat{x}_{k+h^*+1} r_k(h^*+1) = -\begin{bmatrix} \hat{x}_{k+1} & \dots & \hat{x}_{k+h^*} \end{bmatrix} \begin{bmatrix} r_k(1) \\ \vdots \\ r_k(h^*) \end{bmatrix} . \tag{17}$$

From (17) the next value of the state vector can always be uniquely defined by expression

IOP Conf. Series: Materials Science and Engineering **312** (2018) 012025 doi:10.1088/1757-899X/312/1/012025

$$\hat{x}_{k+h^*+1} = -\left[\hat{x}_{k+1} \ \dots \ \hat{x}_{k+h^*}\right] \begin{bmatrix} r_k(1) \\ \vdots \\ r_k(h^*) \end{bmatrix} / \, r_k(h^*+1), \tag{18}$$

where the invertibility of $r_k(h^*+1)$ is guaranteed according to (15) by the condition

$$\begin{bmatrix} x_k & \dots & x_{k+h^f-1} \\ u_k & \dots & u_{k+h^f-1} \end{bmatrix} \begin{bmatrix} r_k(1) \\ \vdots \\ r_k(h) \end{bmatrix} \neq 0. \tag{19}$$

Expression (18) is indeed a universal one-step-ahead prediction problem solution in the absence of errors in the model parameters ($\hat{A} = A, \ \hat{B} = B$) for both non-failure (3) and failure (4) cases.

Then to diagnose the failures we have to make several simple steps.

1. For $k = 1:h$ we need to collect non-failure observations to fulfill the condition (15).
2. For $k = h+1:l$ we predict the non-failure aircraft dynamics $\hat{X}_k$ by (18) with $h^* = h$. Note that in the absence of failures the condition $\hat{x}_k = x_k$ is always satisfied for non-failure case.
3. For $k = l+1$ the predicted state does not equal to real one $x_{l+1} \neq \hat{x}_{l+1}$, so we can conclude that there is a failure in the aircraft control system.
4. For $k = l+1:l+h^f$ we predict the failure aircraft dynamics $\hat{X}_k^f$ by (18) with $h^* = h^f$, and estimate the difference $\Delta X_{k+1}^f = X_{k+1}^f - \hat{X}_{k+1}^f$.
5. The last step is to solve the equation of the form (6):

$$\Delta B U_k = -\Delta X_{k+1}^f. \tag{20}$$

Due to (10) the equation (20) is solvable if and only if

$$\Delta X_{k+1}^f \bar{U}_k^R = 0, \tag{21}$$

and due to (11) the set of all its solutions is defined by a formula

$$\Delta B = \Delta X_{k+1}^f \tilde{U}_k + \Theta \bar{U}_k^L, \tag{22}$$

where Θ is an arbitrary matrix. The fulfillment of the solvability condition (21) is always guaranteed in the absence of disturbances and can be used to control the computations.

From (22) it follows that in order to obtain a unique failures diagnosis problem solution

$$\Delta B = \Delta X_{j+1}^f \tilde{U}_j, \tag{23}$$

it's necessary and sufficient to ensure the linear independence of rows in the control matrix ($\bar{U}_j^L = 0$), that is required for all failures diagnosis methods [18, 19].

Figure 1 shows the implementation scheme of the proposed failures diagnosis method.

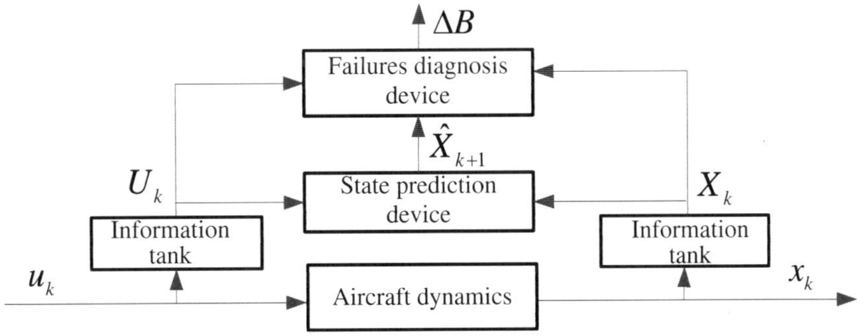

Figure 1. Failures Diagnosis Scheme.

3. Conclusion

The designed method refers to model-based nonparametric methods based on the analysis of input and output signals only and has the following advantages: it doesn't require a priori information about the model parameters, thus there are no any parametric errors; it's applicable for failures diagnosis in unidentifiable dynamic systems; it can reduce the diagnostic time and problem complexity due to no need of identifying the eigen-dynamics parameters; it weakens the requirements to control signals by ensuring only the independence of the controls with no need of exciting all oscillation modes of model.

Acknowledgments

This work is supported by Russian Foundation for Basic Research (17-08-01445-a, 17-01-00251-a)

References

[1] Hajiyev C and Caliskan F 2003 *Fault Diagnosis and Reconfiguration in Flight Control Systems* (Boston: Springer)

[2] Zybin E Yu and Kosyanchuk V V 2010 *Journal of Computer and Systems Sciences International* **49** (1) 105–114

[3] Venkatasubramanian V, Rengaswamy R, Kavuri S and Yin K 2003 *Computers & Chemical Engineering* **27** (3) 293–311

[4] Venkatasubramanian V, Rengaswamy R, Kavuri S and Yin K 2003 *Computers & Chemical Engineering* **27** (3) 313–326

[5] VenkatasubramanianV, Rengaswamy R, Kavuri S and Yin K 2003 *Computers & Chemical Engineering* **27** (3) 327–346

[6] Qi X, Theilliol D, Qi J, Zhang Y, Han J, Song D, Wang L and Xia Y 2013 *Proc. of the Conference on Control and Fault-Tolerant Systems SysTol* 132–139

[7] Blanke M, Kinnaert M, Lunze J and Staroswiecki M 2016 *Diagnosis and Fault-Tolerant Control* (Berlin: Springer)

[8] Zolghadri A, Henry D, Cieslak J, Efimov D and Goupil P F 2016 *Fault Diagnosis and Fault-Tolerant Control and Guidance for Aerospace Vehicles: From Theory to Application* (London: Springer)

[9] Korsun O N, Nikolaev S V and Poplavsky B K 2017 *Mekhatronika, Avtomatizatsiya, Upravlenie* **18**(4) 270–278

[10] Korsun O N, Om M H, Latt K Z and Stulovskii A V 2017 *Procedia Computer Science* **103** 67–74

[11] Zybin E Yu 2015 *Izvestiya SFedU. Engineering sciences* **4** (165) 160–170

[12] Vachtsevanos G, Lewis F, Roemer M, Hess A and Wu B 2006 *Intelligent Fault Diagnosis and Prognosis for Engineering Systems* (Hoboken: Wiley)

[13] Roffel B and Betlem B 2006 *Process Dynamics and Control: Modeling for Control and Prediction* (Padstow: Wiley)
[14] Nava J and Kreinovich V 2015 *Algorithmic Aspects of Analysis, Prediction, and Control in Science and Engineering: An Approach Based on Symmetry and Similarity* (Berlin: Springer)
[15] Mandic D and Chambers J 2001 *Recurrent neural networks for prediction: learning algorithms, architectures, and stability* (NY: Wiley)
[16] Zybin E Y and Kos'yanchuk V V 2016 *Journal of Computer and Systems Sciences International* **55** (4) 546–557
[17] Zybin E, Kosyanchuk V and Karpenko S 2017 *MATEC Web of Conferences* **99** 03011
[18] Isermann R 2006 *Fault-diagnosis Systems: an Introduction from Fault Detection to Fault Tolerance* (Berlin-Heidelberg: Springer)
[19] Ding S 2013 *Model-Based Fault Diagnosis Techniques: Design Schemes, Algorithms and Tools* (London: Springer)